WHITE HOLES

CARLO
ROVELLI

Translated by Simon Carnell

RIVERHEAD BOOKS   NEW YORK   2023

WHITE HOLES

RIVERHEAD BOOKS
An imprint of Penguin Random House LLC
penguinrandomhouse.com

Illustration credits may be found on page 149.

Citations of Dante's *Divine Comedy* are from English translations by
Robin Kirkpatrick, John Ciardi, Henry Wadsworth Longfellow, and the author.

LIBRARY OF CONGRESS CATALOGING-IN-PUBLICATION DATA
Names: Rovelli, Carlo, 1956– author. | Carnell, Simon, 1962– translator.
Title: White holes / Carlo Rovelli ; translated by Simon Carnell.
Other titles: Buchi bianchi. English
Description: First North American edition. | New York : Riverhead Books, 2023. |
"Originally published in Italy as Buchi bianchi by Adelphi Edizioni, Milan, in 2023.
Simultaneously published in Great Britain by Allen Lane,
an imprint of Penguin Random House Ltd., London, in 2023"—copyright page. |
Includes bibliographical references and index.
Identifiers: LCCN 2023015753 (print) | LCCN 2023015754 (ebook) |
ISBN 9780593545447 (hardcover) | ISBN 9780593545461 (ebook)
Subjects: LCSH: White holes (Astronomy)—Popular works. | Black holes
(Astronomy)—Popular works. | Quantum cosmology—Popular works.
Classification: LCC QB991.Q36 R6813 2023 (print) | LCC QB991.Q36 (ebook) |
DDC 523.1—dc23/eng20230722
LC record available at https://lccn.loc.gov/2023015753
LC ebook record available at https://lccn.loc.gov/2023015754

Printed in the United States of America
1st Printing

Book design by Daniel Lagin

*To Francesca, comrade
in science and dreams*

CONTENTS

The most beautiful thing we can experience is the mysterious. It is the source of all true art and science. He to whom this emotion is a stranger . . . is as good as dead: his eyes are closed.

ALBERT EINSTEIN

WHITE HOLES

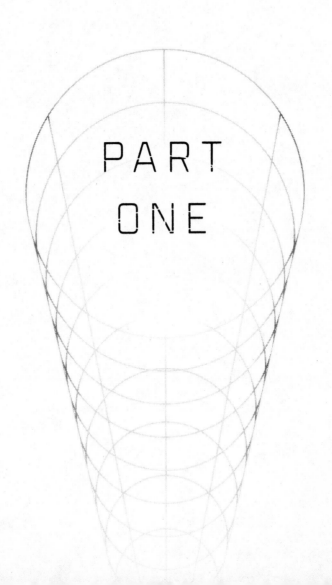

PART
ONE

Beginning is the hardest step. The first words open a space. Like the first glance of the girl I am about to fall in love with—a life is decided by the hint of a smile. I long hesitated before starting to write. Long walks in the woods behind the house, here in Canada. Still not sure where I am going.

For some years now my research has focused on white holes, the elusive younger siblings of black ones. This is my book about white holes. I try to describe *black* holes, which we now see in the heavens in their hundreds, the best I can: what happens at the edge of these strange stars, their *horizon*, where time appears to slow until it stops and space seems to end. And then inside, down, all the way down into the deepest inner regions, where time and space melt. Where time seems to reverse its direction. Where *white* holes are born.

it is a travel tale. it is also the tale of an ongoing adventure. as at the beginning of any journey, i am not sure where it will take me—at that first smile we can't ask where we are going to live together—i have a flight plan in mind. we get to the edge of the horizon. we enter. go down till the very bottom. cross it and break through—like alice through the looking-glass—to emerge on the other side, in a white hole. there we ask what it means for time to be reversed . . . before re-emerging to see the stars again, our familiar stars. after a time that has lasted for just a few seconds—but also millions of years. or maybe for the time that it takes to read this short book.

Are you coming with me?

MARSEILLE. HAL IS IN MY STUDY, STANDING IN FRONT of the blackboard. I am sitting in the big reclining chair, my elbows on the desk, eyes fixed on him. Coming through the window is the clear, dazzling light of the Mediterranean. This is how my adventure with white holes begins.

Hal is American. He has a natural gentleness that tempers the brilliance of his ideas. Today he is a college lecturer; at the time he was still a student. Kind, precise, with the composure and the calmness of someone very mature for his years. He is trying to tell me something I don't quite understand. It is an idea about what could happen to a black hole at the precise moment that its long life comes to an end.

I remember his words: Einstein's equations do not change if we reverse time; to have a rebound, we just reverse time and paste together the solutions. I'm confused.

All of a sudden I see what he means. Wow! (I'm Italian, no calmness.) I go to the blackboard and make a drawing. My heart beats faster.

He thinks about it for a moment. "Yes, that's more or less it." And I: ". . . it's a black hole that quantum tunnels into a white one on the inside—and the outside can stay the same." He thinks about it a bit more: "Yes . . . I don't know. What do you think, might it work?"

It has worked. At least in theory. Nine years have passed since that conversation in the clear light of Marseille. I have continued, with students and colleagues, to work on the hypothesis that black holes can transform into white. It's an idea that seems quite beautiful to me. The idea that I want to tell you about here.

I do not know if it is correct. I do not even know if white holes actually exist. We now know a great deal about black holes—we see them—but no one has seen a white hole. Yet.

When I was studying for my doctorate in Padua, Mario Tonin taught us theoretical physics. He told us he thought that every week the good Lord reads *Physical Review D*, the celebrated physics journal. Whenever He came across an idea He liked, *shazam!* He would put it into practice, rearranging universal laws accordingly.

Hal

If this is so, dear Lord, I would be grateful if you could arrange this: make it so that black holes end up becoming white holes . . .

I READ THE PRECEDING LINES. THE STORY OF MY FIRST encounter with white holes. i want to explain everything in order. what kind of objects hal and i were talking about. what we know, and what we don't know, about them. what the problem was that we were trying to unravel. what hal's idea was exactly, its implications. what it means to reverse time (nothing complicated) and what it means for time to have a direction (perhaps surprisingly, this is the complicated part).

if you follow me, i'll take you to the edge of a black hole's horizon.* we shall enter it, go down to the bottom. we shall cross through and emerge into a white hole where time is reversed. then go up and out until we see the stars again.

let's start the journey: toward white holes.

---

* You may have heard the expression "event horizon." Despite its beauty, I prefer not to use it, because "event horizon" has a technical definition that does not work for black holes turning into white. For a discussion of this technical point, see for instance my scholarly book *General Relativity*.

1

Actually, no, we need to journey toward a black hole first. To understand what a white hole is, we need a clear idea of what a black hole is. What is it?

The first to go wrong was Einstein. In 1915, after a decade of "mad and desperate" work, Albert Einstein published the final equations of his most important theory, one that is studied today in every university in the world: general relativity.

Just weeks after its publication he received a letter from a young colleague, Karl Schwarzschild, at the time a lieutenant in the German army, who in a few months would lose his life as a consequence of the hardship of the Eastern Front.

The letter concludes with these beautiful words: "As you can see, despite the incessant artillery fire, the war has treated me kindly enough to allow me to briefly walk away from all that, and to be able to take this walk on the land of your ideas." A walk on the land of your ideas . . .

Schwarzschild's walk on the land of Einstein's ideas, during a pause in the fighting on the Eastern Front, among the corpses of young Germans and Russians slaughtered by the human stupidity that raged then as now (what can be more stupid than to die for a border?), resulted in an exact solution of the equations that Einstein had just published.

These equations had cost him dearly. We can trace this in their progress through a sequence of articles, each with a different version of them. All wrong.

But in 1915 the equations were finally correct. These are the equations that convinced physicists to revise their ideas on the nature of time and space, showed that clocks go faster in the mountains than in the plains; that the universe is expanding; that there are waves of space—and so on. They are the equations that we use today for studying the cosmos. Perhaps the most beautiful in all physics. (They are the only formula included in my book *Seven Brief Lessons on Physics*.)

In the course of these pages, we will have a close but uneasy relationship with these equations: they will serve as our guide, like Virgil to Dante in *The Divine Comedy*, because they encapsulate the best understanding that we have today of space, time and gravity. They are our tool for understanding. They tell us what to expect at the edge of a black hole and inside it. They also tell us what white holes are. They show us a route through these strange territories.

And yet the whole point of the story I am about to tell is to go and see what happens *where these equations no longer work*. Where it is necessary to abandon them. This is science.

In the middle of our journey we must abandon the reassuring guidance of these equations and be led by something else—just as Dante, in the middle of his journey, leaves Virgil behind and is captivated by something more alluring.

Let's return to Schwarzschild. The solution that he announced in his letter to Einstein is also in all university textbooks today. It describes what happens to space and time around a mass—around the Earth or the Sun, for example. The Earth's and the Sun's gravity causes space and time to bend (I will soon try to explain what this means). It is this curvature of space and time that causes bodies to fall toward the Earth, and the planets to orbit around the Sun. It is the underlying reason of the force of gravity.

The question studied by Schwarzschild was how things are moved by the effect of gravity around something heavy, such as the Earth or the Sun. This is the same question Newton had studied three centuries earlier, opening the path to modern science. Einstein and Schwarzschild correct Newton, improving on his predictions of how things move around masses.

But the solution found by Schwarzschild does not give just some small correction to the movements of the planets. It also predicts something radically new. And strange. If a mass is extremely concentrated, it generates a shell around itself, a spherical surface where everything becomes bizarre: on this surface, clocks—which always slow down in the vicinity of any mass—even come to a stop. Time freezes. It no longer flows. Space, for its part, elongates in the direction of the mass, stretching like a long funnel, and on this absurd spherical surface the stretching becomes a tear. The points immediately within it are already infinitely distant. It's as if space were being pulled apart.

Time stopping, space tearing . . . this sounds bizarre and improbable. Einstein, understandably, concluded that this was nonsense—this absurd surface could not exist in the real world.

In fact, if we do the math, we see that for this surface to be formed it is necessary to crush a mass to a ridiculous extent. To form such a surface around the Earth, for instance, we would need to crush the entire planet down to the size of a ping-pong ball. Absurd! All of this, Einstein concluded, was without interest: you cannot concentrate a mass so much as to arrive at the point where this bizarre shell is formed.

He was wrong. He did not have enough faith in his own equations. He did not have the courage to believe the strange implications of his own theory. We now know better: masses as concentrated as this do exist. As a matter of fact, there are billions upon billions of them in the heavens. They are what we call black holes.

Astronomers have observed them, varying in size from a few kilometers in diameter to truly colossal, as big as the entire solar system, or bigger. There may also be small ones (as small as a ping-pong ball) and even extremely small ones (weighing no more than a hair), but for now we have not seen them. For now.

Most of the black holes identified in the sky originated from stars that finished burning. They were large stars, so heavy that they would have collapsed on themselves had it not been for the fact that they were burning. Stars burn the hydrogen of which they are made, transforming it into helium. The heat produced by this combustion generates a pressure that counterbalances the weight of the star, preventing it from being crushed by its own weight. In this way a star continues to live and burn for billions of years.

But nothing is forever. In the end, the hydrogen in the star is consumed and turned into helium and other ashes that burn no more, leaving the star like a car without fuel. The star's temperature drops, its weight begins to prevail. The star is crushed by the effect of gravity. The gravitational force in a large star is tremendous—not even the hardest rock could withstand its pressure—now that there is no longer anything to prevent the star from collapsing in on itself. It collapses, all the way down to its horizon. A black hole is formed.

IN 1928, LONG BEFORE THIS PROCESS WAS UNDERSTOOD, the Bell telephone company hired a physicist, Karl Jansky, to study the noises that disturb radio communications. Jansky built a rudimentary thirty-meter-long directional antenna: a bizarre grid of metal rods mounted on wheels that could rotate in any direction. His colleagues dubbed it "Jansky's merry-go-round." Here it is:

With this antenna, Jansky started to record all radio signals he could find: flashes of passing thunderstorms, noises caused by radio antennae and so on. Among the others, a curiously regular signal was detected, a kind of hiss captured at every turn of the merry-go-round.

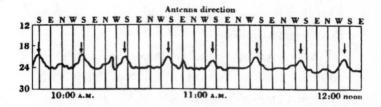

Jansky's sister tells how their father raised his children encouraging them to "investigate everything." Jansky investigated the "hiss" for more than a year. The hiss increased and decreased in intensity every twenty-four hours. Jansky thought it could come from the Sun, which hovers over us once a day. But the devil is in the details. Continuing to monitor the signal, Jansky noticed that the period was not twenty-four hours but a little less: twenty-three hours and fifty-six minutes. That is, the stronger signal was not always at the same time of the day. It kept sliding earlier, like a clock running a bit too fast. This was strange. It could not be the Sun: the Sun does not run faster than the days.

Stars do. A fellow astronomer pointed out that twenty-three hours and fifty-six minutes is exactly the period it takes for the *stars* to rotate around us. (The stars take a little less time than the Sun to rotate in the sky, because Earth and Sun dance around each other with one waltzing turn per year.) The mysterious radio signal can only be coming from the stars! The direction was easy to find: it originates from the stars toward which the antenna is pointed when the signal is strongest. By consulting a star atlas, it must be coming from the center of our galaxy . . .

This was such sensational news that it ended up in *The New York Times*, under the headline "New Radio Waves Traced to Centre of the Milky Way." On May 15, 1933, with millions of Americans listening, NBC broadcast live the hiss coming from the stars, with an interview of Jansky. "Good evening, ladies and gentlemen, tonight we are listening live to radio signals received from outside the solar system, from somewhere among the stars." Jansky explained to the audience that the signal was coming from the center of the galaxy. The announcer commented that the strength of a signal reaching us from thirty thousand light-years away must be "immense," and that it must be "millions of millions of times more powerful than any radio station on earth."

Five days earlier, in the Opernplatz in Berlin, the Nazis had staged the biggest of their book burnings. Among the texts burned were works by Vladimir Mayakovsky ("My verse will reach you . . . not as the light from dead stars reaches you"[1]) and books by and about Albert Einstein. Ninety years later, thanks to the ideas in those books, we know what the mysterious whistle heard by millions of Americans is. It is the radiation emitted by incandescent matter that whirls furiously before falling into a colossal black hole at the center of our galaxy. A black hole ten times as big as the entire orbit of the Moon around the Earth, with a mass four million times that of our Sun.

i am on my third revision of these pages, and today the astronomers have released an image of precisely this black hole. the image shows the burning material that whirls around the hole, generating the same radiation captured a century ago by jansky's antenna. here it is:

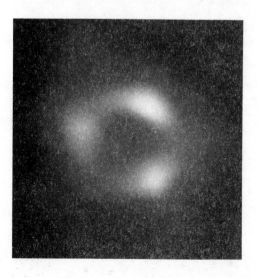

i am moved intensely by this image. i've studied black holes all my life, without knowing if they truly existed . . . and now here they are. here is the visual evidence, something i could never have imagined when i was a university student, charmed by these holes.

Twenty years ago there were still many who doubted the existence of black holes. When in January 2000 I moved from the United States to France, my new department head asked me, "You don't really believe that black holes actually exist, do you?" Now he has changed his mind. This is not meant as a criticism; it illustrates, rather, the beauty of science: there is nothing wrong with changing your mind, given that we are all still learning. The best scientists are the ones who change their minds often, as Einstein did.

In the image above, the actual black hole, or rather the *horizon*—as we call the bizarre surface enclosing the black hole—is a small, dark disk in the center, in the heart of the fiery matter that revolves around it.

This horizon is our way in.

2

We approach the threshold, the horizon. What happens at the horizon of a massive black hole, past the material that swirls so violently and becomes so incandescent that it can be detected by a grid of iron rods thirty million light-years away?

It took decades to understand what happens on the horizon. Einstein was not the only one to have understood nothing. Physicists remained confused for a long time. Only in the second half of the twentieth century did we begin to understand horizons. There are quite a few colleagues in my field who get confused about them even today (and this *is* meant as a criticism).

Let's go and see.

In a work called *The Dream*, Johannes Kepler, the seventeenth-century astronomer who first understood how the planets turn around the Sun, describes how his mother took him around the solar system, as on a flying broomstick, to show him the solar system from a different perspective.

Kepler's mother was tried for witchcraft. Just in case you are wondering if she really was a witch: at her trial, defended by her son, she was acquitted.

Kepler, perhaps thanks to his mother, the witch, wanted to go and see. To go and see: this is science. To go and look where we have not been before, by using logic, reason, math, intuition, imagination. Around the solar system, to the heart of atoms, inside living cells, within the convolutions of the neurons of our brain, far in the past, beyond the horizon of black holes . . . To go and see with the eyes of the mind.

ON EARTH WE CALL "HORIZON" THE DISTANT LINE beyond which we cannot see. If we get on a ship and sail toward it, we can cross that line: we can go *beyond* the horizon. When we cross it, nothing particular happens. Except that we vanish from the view of anyone watching us from the shore, without this involving anything special happening to the ship (perhaps a party on board, in some maritime traditions).

And here's the surprising thing: the same is true for the

"horizon" of a black hole. Traveling in a spaceship we can get as close to the horizon as we like. We reach it. We cross it. Nothing special happens. Our watches continue ticking at their normal rate, the distances around us continue to be the same.

Only, as for the ship on the sea, we can no longer be seen by watchers from afar. We are beyond *their* horizon. If we try to signal by sending back, toward the outside, a ray of light, the ray cannot escape. It is trapped within the shell of the horizon. We can no longer reach our distant friends. Inside the horizon, the force of gravity is so strong that it traps everything, even light.

⤙⬩⤚

WHY, THEN, DOES THE SOLUTION FOUND BY SCHWARZ-schild indicate that at the horizon clocks stop and space breaks down, confounding Einstein and all the others? If you can cross the horizon and find everything normal there, is Schwarzschild's solution wrong?

It is not wrong. It is just written from the perspective of someone distant from the horizon. Schwarzschild's solution is like a geographical map of the space *outside the horizon*.

With geographical maps—and this is well known—something peculiar happens. Let's take a map of the earth made up of two disks.

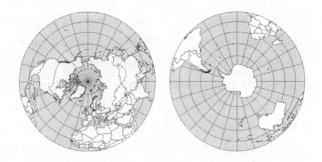

The equator seems very strange in this map. It seems to be the edge of the world. In reality it is obviously not the world's rim, and nothing special happens there (apart from the heat). The surface of the earth is not flat, so it does not fit well into just one map—but it does not finish at the edge of the map. Spacetime is not flat either, and similarly it does not fit into a single map—but it does not end at the edge of Schwarzschild's solution either.

This is what happened to Einstein and all the others: they interpreted the solution badly, like someone looking at the left map in the figure above and deducing that the earth ends at the equator. This mistake was made by dozens of outstanding scientists for decades (and there are scientists, even among the big shots, who make it still).

How do we know that it is a mistake? None of us has actually gone yet to witness in person what happens at the horizon of a black hole . . .

None of us has gone there, but we have the theory. The

same set of equations that yields the Schwarzschild solution also allows us to compute what happens when we approach the horizon. Working this out is not even that difficult. I set it as an exercise for my students when I teach general relativity. But it took time for someone to think about doing it in the first place, and to understand what it meant.

This someone was David Finkelstein, in 1958 (I was two years old). Finkelstein was a vastly cultured scientist, with interests that included politics, art and music as well as science. He had a capacity for deep and daring thinking. He left

David Finkelstein

us not long ago, in 2016. I was lucky enough to meet him, in the last years of his life, with his long prophet's beard and a manner that was somehow both laid back and hieratic. He was one of those rare scientists who open new vistas or pathways for thought. We will encounter him again later in this story.

In 1958, Finkelstein published a beautiful, brilliant article in which he clarified the nature of the horizon. The title of the article is "Past-Future Asymmetry of the Gravitational Field of a Point Particle."[2] It sounds technical, but keep this title in mind, because the idea it expresses will be the keystone to our story: "Past-Future Asymmetry."

Finkelstein's calculation shows that if we approach the horizon and go beyond it, our watches *do not* slow and nothing strange happens to the space around us, just as nothing peculiar happens to a ship when it crosses the line of the horizon and disappears from our view.

SO WHY DO WATCHES APPEAR TO STOP IN SCHWARZ-schild's solution?

Because Schwarzschild's solution describes what happens there—*as seen from afar.* From afar, all timepieces *appear* to slow down and stop as we reach the horizon. And there is no contradiction between these two perspectives.

Let's imagine that we are traveling through countries

where the mail gets gradually slower, and that we are sending a letter home to our father every day. Our father will receive the letters at lengthening intervals, because we will be arriving in places where the postal service takes longer to forward them. For him it will seem as if we have slowed down: initially he will get news of our day from us every day—but then he will have to wait several days, and eventually even weeks, just to learn about a new single day of ours. For him it is as if our life has slowed . . .

If we reach the desert, where there is no mail at all, he will receive only the last letter we wrote before entering the desert, which will arrive a long time after it was sent. For our father, the edge of the desert is therefore the place where *for him* our time stops. It is the horizon beyond which we can no longer be seen *by him*. He will continue to visualize us frozen at the desert's edge.

Something similar happens if we cross the horizon of a black hole. If our father looks at us going toward the horizon, he sees the ticking of our watch getting gradually slower, because as we get closer the light takes longer to move away from us and reach him. The light lingers near the horizon, held back by gravity, before being able to escape. If our father continues watching, he will see the moments of our life gradually slowing until they come to a halt—frozen with the final tick before we cross over.

In the desert, or beyond a black hole's horizon, we

continue to exist as normal. But our father will no longer receive anything from us, however long he waits.

Time, in short, does not come to a stop *for those who have crossed* the horizon. It is only for those looking from afar that what happens near the horizon appears dramatically slowed down.

THE ANALOGY WITH LETTERS SENT WHILE APPROACH-ing the desert is a good one. But only in part. There is a difference between the trip to the desert and the trip to the horizon. If instead of going all the way into the desert we go back and embrace our father again, the hours that will have passed since we last saw him will be the same for us both. If he is a year older, we will also be a year older.

The same does not hold true for the temporal distortions in the vicinity of the black hole's horizon. These are real and concrete, in the following sense. If we approach the horizon, linger in its vicinity and then turn back, the time elapsed for us between last seeing him and embracing him again will be *shorter* than the time passed for him. He will have aged more than we have.

This is not a perspective effect; it is a real distortion of time, due to gravity. Where gravity is strong, time actually passes more slowly than where it is weak, in *this* sense. Time effectively passes at different rates in different places.

This is what we mean by saying that spacetime is "curved." There is less of it near the horizon, and this fact "bends" spacetime, the way taking away some stitching bends knitted fabric.

IN SUM, CLOSE TO THE HORIZON, TIME SLOWS IN THE sense that anyone observing us from afar would see us moving in slow motion, and *also* in the sense that if we turn back, more time will have passed for those who remain at a distance than it has for us. In another sense, on the other hand, time does not slow: for us who are at the horizon, no slowing is felt. For us, time passes normally.

Perhaps, dear reader, it has occurred to you at this point to ask which of these different times is the "real" one: the time of those at the horizon, or the time of those watching from a distance? The answer is neither of them. The revolution brought about by Einstein was the realization that the question of which time is the true time . . . is meaningless. It is like asking which regions of the earth are truly "above" and which are truly "below." With respect to each location, this location is above, and the others are below. Every place on earth determines a different "above" and "below." The difference is only a question of perspective. In the same manner, every place in the universe has its own time. Different places can send each other signals—like the whistle sent to us from the black hole at the center of our galaxy—but time passes at unequal rates in different places, and no single one of these times is "truer" than any other.

In other words, the slowing-down of time in the vicinity of the horizon is something that regards only the *relation* between how time passes in different places. It is only in relation to the time of a distant observer that time stops or is slowed down at the horizon.

The world is woven by these *relations* between times. There is no universal time. Temporality is the network formed by the many local times and the possibility of exchanging signals. From close up, the horizon is a normal place. From a distance, it is the place where time stops.

This is what David Finkelstein understood.

FINKELSTEIN WROTE AN ARTICLE[3] ABOUT AN ENGRAVing by the renowned Renaissance master of perspective Albrecht Dürer. The engraving, titled *Melencolia I*, is a complex work, charged with symbolism.

I think it is no coincidence that the first person to understand the horizon of black holes was not a mathematical genius, but someone with the capacity to write about Albrecht Dürer and Renaissance perspective.

The Renaissance discovered the art of perspective in paintings. It discovered that our entire access to reality is

perspectival. According to Finkelstein, the ambiguity of the engraving reflects and represents the unavoidable ambiguity between perspectives. In his reading, the engraving stages the melancholy of those who vainly attempt to reach absolute truth and beauty. The impossibility of obtaining the absolute—for Dürer and Finkelstein—is the source of our melancholy.

it is not the case for me. on the contrary, it seems to me to be the source of a gentle dizziness. the dizziness of lightness, of the intoxicating inconsistency of that delicate texture that is the reality of which we are part.

we have access only to perspectives. reality is perhaps nothing other than perspectives. there is no absolute. we are limited, impermanent. and precisely for this reason, to live, to be, as we do, is so light and sweet . . .

3

We are about to cross the horizon and observe the black hole from the inside. Before going in, however, allow me a detour. You can skip ahead, if you are a hasty reader.

As yet, we have only approached the horizon; we have not even entered. And yet we have already encountered something disconcerting: the relativity of time. The relativity of time is an established fact, but a fact still difficult to digest—perhaps the most difficult of all—for the journey we are traveling.

Dante, too, encountered the greatest difficulty, in the form of three fierce wild beasts, before he had even crossed the fateful threshold of the Inferno. Like any traveler, he knew that the first step, abandoning the familiar paths, is the most difficult.

How did such a bizarre idea as the relativity of time come to be devised and accepted?

Conceptual leaps like this one are hardly unique to contemporary science. On the contrary, they form a deep current that has always nourished the growth of our understanding of reality. It is how we learn in earnest: by changing some of the basic ideas that seem self-evident to us.

We figured out (two millennia ago) that the Earth is spherical, and (half a millennium ago) that it moves. At first glance these are absurd ideas, since the Earth appears to us to be flat and still. In order to digest such ideas, the difficulty lies not so much with the new concept as it does with becoming liberated from old ones that seem so obviously to be true; bringing them into doubt seems inconceivable. We are always convinced that our natural intuitions are self-evidently right, and it is this that prevents us from learning more.

The difficulty lies not in learning, but in unlearning. In Galileo's great book the *Dialogue Concerning the Two Chief World Systems*, most of the text is not dedicated to arguing that the Earth turns. It is dedicated to demolishing the ingrained intuition that its turning is inconceivable.*

Arriving at the relativity of time took twenty-six centuries of similar conceptual leaps. I'll summarize them here in

---

* In fact, after the masterful demolition of the idea that Earth's stillness is necessary, Galileo relegates the demonstration that the Earth actually moves to the last "day" of the book, and his demonstration . . . is wrong.

a rapid bird's-eye view of two and a half thousand years of thought:

1. *Anaximander* (sixth century BCE) comes first. If Sun, Moon and stars revolve around us, then there must be empty space below the Earth as well as above it. *The Earth, therefore, hovers upon the void.*

2. *Aristotle* (fourth century BCE) observes that eclipses of the Moon show how the disk of the Moon is only slightly smaller than the disk of the Earth's shadow. Hence *the Moon is a large heavenly body*, only a little smaller than the Earth.

3. *Aristarchus* (third century BCE) notices that when the Moon is a quarter, the angle between the Sun and the Moon in the sky (α in the figure below) is almost a right angle (do try to measure it when the Moon is a quarter: it's easy). The triangle Sun-Earth-Moon therefore has two angles that are almost right angles (as the Moon is half lit).

But a triangle with two angles close to ninety degrees has a very distant vertex. Hence the Sun is much more distant than the Moon. Yet the Sun and Moon appear to be equally large in the sky. So the Sun must be much larger than the Moon—which implies that *the Sun is gigantic, much larger than the Earth!* And so it is reasonable to think—Aristarchus suggested, twenty-three centuries ago—that it is the little Earth that dances around the gigantic Sun, and not the other way around.

4. We have to wait for *Copernicus* (sixteenth century) and *Kepler* (seventeenth century) for this line of thinking to show its power in accounting for the movements of the planets, and for the persuasive rhetoric of *Galileo* in his *Dialogue* (also in the seventeenth century) to convince humanity that it is actually *the Earth that moves*, contrary to our intuition.

5. Building on the results of Copernicus, Kepler and Galileo, in the eighteenth century, the greatest scientist of all, *Newton*, constructs modern physics. He asks what it is that holds the Earth and other planets in their orbits. He imagines that all objects have a "natural" motion (an idea of Aristotle's), at a constant speed (an idea of Galileo's), *in a physical space described by Euclidean geometry* (an idea of his own), but they are deflected by "forces." With

superb mathematical skill, he demonstrates that *the force that keeps the planets and the Moon in their orbits is the same as the familiar "gravity" that draws us downward.* The idea of a "force" that acts at a distance is Newton's stroke of genius. It is the intuition of the existence of something else in the physical world, beyond material bodies that hurtle and collide.

6. Studying the forces of electricity and magnetism, *Faraday* and *Maxwell* (nineteenth century) understand that forces are not instantaneous. There is a lag between cause and effect: the time taken for light to travel. Light is fast, the interval short. Newton was *almost* right: the effect is *almost* instantaneous. But not quite. *"Something" that is spread throughout space gradually transports the force from one body to another.* We call this "something," intuited by Faraday, the "physical field." Electric, magnetic and gravitational fields transmit the forces. Maxwell writes the equations for the electric and magnetic fields.

7. While searching for the corresponding equations for the gravitational field—the equations to which Schwarzschild found a solution—*Einstein* (twentieth century) stumbles upon the most spectacular discovery since Anaximander realized that the Earth hovers upon the void without the need to lean on anything. He realizes that *the geometry*

of time and space, measured by rulers and clocks, *is determined by the gravitational field*—the field that carries the force of gravity. The equations for the gravitational field *also* describe (it's the same thing) how space and time are distorted. *This, then, is gravity: a distortion of time and of space*, influenced by things. The distortion includes the slowing of clocks I have described above.

This is how humankind has arrived at understanding the distortion of time.

The mass of the Earth slows time in its vicinity. This is by a small amount, but it can be measured with precision timepieces. Its most conspicuous effect is the gravity with which we are most familiar, which makes heavy things fall. Falling is therefore a direct consequence of the slowing of time. It would take a bit of math to show in detail how, but *a stone falls precisely because it follows a straight trajectory in spacetime distorted by a local slowing-down of time.*

This amazing idea—that gravity is the effect of the distortion of space and time—is Einstein's theory of general relativity. It brings into doubt something that seemed self-evident to us: that the geometry of physical space must be the geometry of Euclid that we studied at school, and that time passes everywhere at an equal rate. It is an extremely

simple idea (like Anaximander's) and a disconcerting one
(like Anaximander's).

End of detour.*

---

* If this extremely condensed account of the entire history of theoretical physics
turns out to be incomprehensible, no matter: it is not needed for what follows.
If you find this tale interesting, you can read about it in detail in *Reality Is Not
What It Seems*.

4

So here we go. We are at the edge of the horizon. Let's cross it. Thanks to Finkelstein, we are not afraid that the world will end here. It is not the first time that dark advice—*Abandon all hope, ye who enter here**—turns out to be unduly threatening.

We enter, then, with the courage of those who launch themselves toward the unknown. With the voice of Ulysses in our ear: *Do not pass up the chance to experience, beyond the sun, the world where no one lives. Consider your origin: you were not made to live like brutes, but to pursue virtue and*

---

* Dante Alighieri, *Divina Commedia*, Canto III of the first section, the *Inferno* (Hell). The quote is the inscription on the gate of Hell. It threatens whoever enters, the way many textbooks today condemn whoever enters a black hole horizon to never again see the light. Dante nevertheless stepped through the gate, and after a long trip he came out alive. So, do not worry: we too will come back alive—via a white hole.

knowledge.* And like the companions of Ulysses, *on this mad flight we'll make wings of our oars.*

We are now within the black hole, *into the secret things.*†

With a good map of the stars (which remain visible from the inside), we are able to recognize that we have crossed the threshold beyond which we can no longer send letters home. It is too late now to stop and turn back. Beyond the horizon, where not even light can escape, we have even less chance than light of doing so. However many powerful rocket engines we might have taken with us, there is now no way of avoiding a fall toward the center.

In order to come out again, we *must take another path.*‡

With a little attention, we can also account for the fact that we are *inside* a black hole just by looking around. Here space is spherical, just like it is outside, around the horizon— but outside, with powerful enough rockets, we can move (upward) toward larger spheres. Inside, on the contrary, whatever we do we will find ourselves in ever smaller spheres. The gravity that pulls us down is so strong that we can do nothing to impede our descent.

---

* *Inferno*, XXVI. Dante meets Ulysses. These are the words with which Ulysses says he addressed his companions, to motivate them to start a new adventure.

† *Inferno*, III. So Dante calls Hell as he enters.

‡ *Inferno*, II. These are the words Virgil uses to tell Dante that he will have to take a long path (through Hell, Purgatory and Paradise) to be able to come back.

Like Dante and Virgil in the circles of the Inferno, we go down.

THE GEOMETRY OF THE SPACE INSIDE A BLACK HOLE, *down there in the blind world below*,* is actually surprisingly similar to that of Dante's Inferno. Think of a funnel. A very long funnel. At any given moment, the interior of the black hole can be imagined as this funnel.[4] The older the black hole, the more elongated its interior. The interior of a very old black hole might even be millions of light-years in length. The image below illustrates how we can think of the inside of a black hole at a given moment.[5]

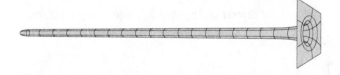

However enormous it might be, the length of the funnel is not infinite: at the bottom there is still the star that, by collapsing in on itself, gave rise to the hole.

---

* *Inferno*, IV. So Virgil calls the Hell where they are entering.

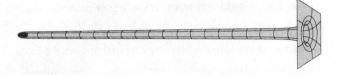

Unlike Dante's Inferno, however, which as far as we know stays the same, the funnel here *lengthens and narrows* with the passage of time.

In order to illustrate this fact, the drawing below shows a succession of funnels, each representing the interior of the black hole at a successive moment. Following the habit of physicists, the drawing shows a sequence in which time runs from the bottom to the top of the image. I do not know why this is the convention (perhaps we have borrowed it from geologists, who always draw the past below, because the oldest layers are deepest underground). The drawing must there-

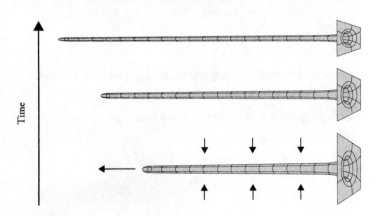

fore be read from the bottom upward: going up, the tube lengthens and narrows.

If we go down into the black hole, at each moment we are at a point in this funnel, going farther and farther down:

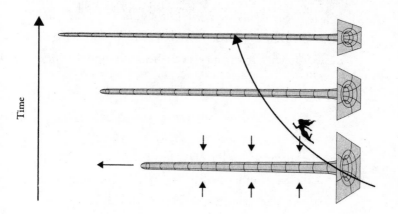

This is the shape of space inside a black hole: an endless abyss (*dark and deep and nebulous it was**) that gradually constricts around us as we fall, without our being able to reach the bottom, where the star that gave rise to it in the first place has fallen.

How do we know all this, about the shape of the interior of a black hole, if no one has been there to see it and returned to tell the tale? We know it because the inside of a black hole

---

* *Inferno*, IV.

is fully accounted for by Einstein's equations. Until something happens to create any doubt, we have no reason not to trust in these equations, since their predictions—spectacular and unexpected as they were—have so far *all* turned out to be correct.

These equations are our faithful guide: like the good Virgil for Dante—*Duke, Lord and Master**—showing us the way down, farther and ever farther down into the *blind world*.[†]

---

* *Inferno*, II.
† *Inferno*, IV.

5

Sooner or later, however, even the best guide no longer suffices. Sooner or later something always happens that causes us to doubt them. Or as Linji Yixuan, great Chinese master of Buddhism in the Chan tradition has it, "If you meet the Buddha on the road, kill him."[6]

Deep down where we are falling, there are regions where the distortion of spacetime is extremely strong. Here, we expect quantum effects to intervene, as always happens in extreme conditions. Einstein's equations do not take account of such phenomena. They ignore them. In these regions, therefore, those equations no longer serve. We have lost our guide.

In effect, we are certain that at some point Einstein's equations are no longer good, because if we continue to rely on them, they go mad. They predict that the geometry will reach infinite distortion, and here they no longer work: the value of the variables in the equation becomes infinite—and we cannot go any further. Einstein's theory, our staunch guide, does not help us anymore. We call these regions—with their spikes, cusps and folds—"singularities."

But the devil is in the details. Let's see exactly *where* the equations stop working. Pay attention. This is the detail that has caused the most confusion among scientists. It still confuses many, even some of the best. It was clarity on this detail that allowed Hal and me to escape the impasse.

It may seem natural to think that strange things happen *at the bottom of the funnel*, down in the *center* of the black hole, in the dark area of the image.

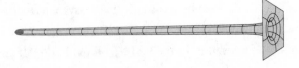

But this is not the case. In the center of the funnel there is only the falling star; we are not in singularity territory. Here, the equations still work.

How so? If we go into a black hole that is very old, has the star not finished falling for a good while? Has a long time not passed since it collapsed? A star that collapses on itself is crushed to a point in a very short lapse of time. So how come it is still there, still in the process of falling, after so long?

Time . . . It is always the crux of the matter. "Long time" for one does not mean "long time" for another. "Long time" for us does not mean "long time" for the star. Down there, at the bottom, time has slowed tremendously. Outside, millions of years may have passed, while down there just a few fractions of a second . . . The star is still falling at the bottom of the long funnel that is stretching and narrowing,[7] because in *its* time no more than fractions of a second have passed. So the zone where the distortions become infinite, where the equations of Einstein stop working, the interesting zone . . . is not there!

So where is it?

It is *in the future*. It is in what happens *after* the interval of time depicted in the last image. It is in the gray area of the image below.

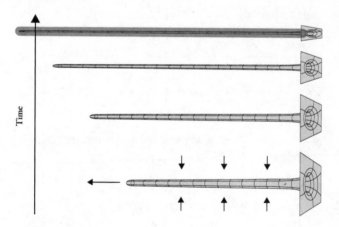

As the diameter of the funnel narrows, the cylinder becomes more curved, like a roll that is rolled tighter. The more the funnel is narrowed, the greater the distortion of spacetime. When this reaches the fateful "Planck scale,"[8] the scale where we expect space and time to be seriously affected by quantum phenomena, we enter the region where these phenomena imply that Einstein's equations are violated.* This is the gray zone in the image above.

If we ignore these phenomena and trust in Einstein's theory, the equations predict that the crushing of space continues until catastrophe occurs: the long, thin tube becomes longer and thinner until it is squeezed into a single line (crushing us with it).

And then? Then that's it. Space has collapsed, time is finished. We hit a wall. If we go along with Einstein's theory, time ends here.

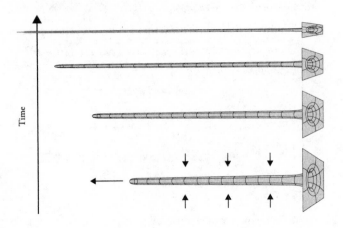

---

* Quantum phenomena are characterized by a constant, the Planck constant, that determines this scale.

So the region of the singularity, the quantum region, *is in the future*, where the tube is squeezed into a line and becomes infinitely long. It is not *in the center* of the ball that is the black hole—as, alas, many continue to think. There is only the falling star in the center. This misunderstanding is a main source of the confusion about the fate of black holes.

To understand what happens in a black hole, in other words, we should not think of a stationary cone with the singularity at its center. We need to think of it as a long tube with the star that generated it at the bottom: the tube gets longer and thinner and *in the future* is squeezed into a line. The singularity is not *at the center*; it is *after*. This is the key to the story.

Falling into the black hole, that is where we end up. *That is the last depth and the darkest lair. And the farthest from Heaven which encircles all.**

---

* *Inferno*, IX. This is what Dante calls the deepest circle of Hell.

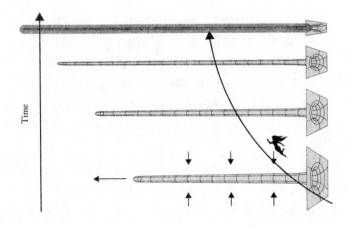

With that, we have reached the quantum zone. So what happens now?

Einstein's equations—our guide—the most beautiful equations in all physics, the ones that have accompanied me throughout my scientific life: they are no longer enough. Now we have no guide, just as Virgil left Dante: *He had gone. Virgil had gone. Virgil the gentle Father to whom I gave my soul for its salvation.**

What happens next? This is what Hal and I were discussing on that clear afternoon in Marseille.

---

* *Purgatory*, XXX. Dante loses his guide after having traversed Hell and Purgatory.

6

How can we proceed when our guides no longer suffice? Without the stars it is perhaps more beautiful to navigate—but how do we learn something new, something we do not yet know?

To learn something new, one way is to go and experience. Over the next hill. This is why the young depart and travel. Or someone might have gone there for us. What they have learned comes to us as a story, a lesson at school, a Wikipedia entry, a book. Aristotle and Theophrastus go to the island of Lesbos, they minutely observe fish, mollusks, birds, mammals and plants—they write it all down in books and open up the world of biology.

To see still farther, there are tools. Galileo points a telescope at the heavens and sees things we people wouldn't believe, opening our eyes to the unending vastness of astronomy.

Physicists use spectrometers to analyze the light emitted by elements and collect data on the atoms, opening the door to the quantum world. Accurate observations with instruments are at the root of so much new knowledge. But we can neither reach nor observe anything at the bottom of a black hole if no light can escape from it . . .

If we cannot travel there physically, however, we can do so mentally. We can *imagine* changing perspective, to see things in a different way.

Anaximander, the first pioneer of science on the list in chapter 3, was remembered in antiquity as the first to have designed geographical maps. Such a map is a bird's-eye image of a large area as it would look if we could fly higher than eagles. For millennia of civilization, travel and trade, no one had come up with this idea. It was not an easy leap. We are habituated to seeing the Earth close up: who had ever seen it from so high? To identify with an eagle, to wonder what it would see from such a great height—this is to change perspective. Anaximander had the imagination to make such a leap. He also had the courage to imagine what the Earth itself would look like from an immense height. It is in this way that he could intuit how Earth would appear to Armstrong and Aldrin as they looked down at it from the Moon.

The greatest astronomer of the ancient world was Hipparchus. A result attributed to him beautifully illustrates the value of traveling elsewhere with the mind. It is his calculation on the distance to the Moon. It is summarized in the following figure (out of scale, the Sun being much bigger and much farther away) and in its long caption.

The first step in Hipparchus's refined geometrical argument is the question "What would I see if I went to the tip of the Earth's shadow cone?" Imagine yourself down there, thousands of miles from the Earth, in interplanetary space,

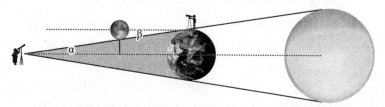

Hipparchus imagines flying to the tip of the cone made by the Earth's shadow, then looking back. Looked at from there, the Earth hides the Sun perfectly. Hence the angle α is half the angle within which we see the Sun. The angle β is half the angle under which we see the Moon. But Sun and Moon appear to be of equal size in the sky, hence α = β. Euclid's geometry informs us, then, that the two dotted lines are parallel, and the drawing shows that the radius of the Moon plus the radius of the shadow (where the Moon is) makes a segment equal to the radius of the Earth. Observing an eclipse shows that the radius of the disk of the shadow is two and a half times the radius of the Moon, hence the radius of the Earth is three and a half times the radius of the Moon. A coin 1 centimeter in diameter covers the Moon if we hold it 110 centimeters from our eye (try it for yourself!), hence the distance to the Moon is 110 times its diameter. Hence, the distance to the Moon is 110 divided by 3½—that is, around thirty times the diameter of the Earth. And this is exactly right! Pure genius. All this from simple observations that we can make in our garden with the naked eye.

looking back and watching the Earth cover the Sun ... seeing with the mind's eye.

Copernicus looks at the solar system as you would see it from the Sun. Kepler flies thanks to his mother's magic, and describes the solar system as seen from outside the Earth. Einstein wonders what he would see if he could ride a ray of light. We project ourselves into situations ever further from our everyday experience. We imagine looking at everything from a different perspective. We ask what we would see if we could go into a black hole.

BUT HOW DO WE "GO AND SEE" WITH THE MIND'S EYE? Anaximander did *not* soar with eagles, Kepler did *not* (no, really, he didn't) fly to the Moon on a broomstick, and Einstein did not ride a ray of light. How can we go and see from somewhere that we cannot actually reach?

I think that the answer is to grope for a delicate balance. A balance between how much we take with us and how much we leave at home. What we carry with us allows us to know what to expect. To know what to expect in the black hole, we have used the equations of Einstein, which predict its geometry. Einstein used the equations of Maxwell; Kepler used the book of Copernicus. These are the maps, the rules, the generalities that we trust in because they have worked so well.

And yet, we know that we must leave something behind.

Anaximander left at home the idea that all things fall in parallel; Kepler the idea that things move in circles; Einstein the idea that all clocks tick in tune with each other. If we leave too many things at home, we lack the tools needed to forge ahead; if we take too many, we fail to find the paths to new understanding. There are no recipes for success; there is only trial and error. *Trying and trying again.** This is what we do . . . *the long study and the great love.*†

We combine and recombine in different ways what we know, looking for a combination that clarifies something. We leave out pieces that previously seemed essential, if they get in the way. We take risks, albeit calculated ones. We linger at the border of our knowledge. We familiarize ourselves with it, and we spend a long time there, walking back and forth along its length, searching for the gap. We try out new combinations. New concepts.

Our new concepts are old ones adapted and modified. We think through analogies. Newton's "forces" are borrowed from the everyday experience of pushing and pulling. Faraday's

---

* *Paradise*, III. Dante writes that he learns new truths from Beatrice "provando e riprovando." In modern Italian this expression means "trying and trying again," and it is often quoted in this sense. In thirteenth-century Italian it more likely meant "by proof and refutation." The language has perhaps evolved in parallel with the epistemology.

† *Inferno*, I. In meeting Virgil for the first time, Dante uses these words to describe his relationship with the text of the great Latin poet. Or maybe the relation of any intellectual with knowledge, including the relation of any scientist with their topic.

electric and magnetic "fields," extended in space, are stolen from farmers. Einstein understood that time passes sometimes slowly, and sometimes fast—but didn't we know that already?

The West has been able to use the creativity of analogical thinking very effectively, building new concepts generation upon generation, building the castle of scientific thought, a magnificent legacy to today's global civilization. But the fact that thought grows and develops analogically rather than syllogistically was most clearly recognized in the East. The logic of reasoning based upon analogies had been analyzed as early as the Mohist school,[9] and is implicit in one of the greatest books ever written, the *Zhuangzi*. Scientific thinking makes good use of logical and mathematical rigidity, but this is only one of the two legs upon which its success has been built. The other is the creative liberty taken with conceptual structure, and *this* grows through analogy and recombination.

An electromagnetic field is not a field of wheat; Einstein's slowing-down of time is not the one that we experience when bored; there is nobody pushing and pulling where Newton's forces act. But the analogies are manifest. Making an analogy involves taking an aspect of a concept and using it in another context, preserving something of its original meaning while letting something else go, in such a way that the resulting combination produces new and effective meaning. This is how the best science works.

I think that this is also how the best art works. Science and art are about the continual reorganization of our conceptual space, of what we call meaning. What happens when we react to a work of art is not happening in the art object itself, of course—still less to some mysterious world of the spirit: it lies in the complexity of our brain, in the kaleidoscopic network of analogical relationships with which our neurons weave what we call meaning. We are involved, engaged—for this takes us out of our habitual sleepwalking, reconnecting us instead with the joy of seeing something anew in the world. It is the same joy that science gives. The light in Vermeer's painting opens our eyes to a resonance of light in the world that we had not previously been able to seize; a fragment of a poem by Sappho (*bitter-sweet is Eros*) opens up a world in which to rethink desire; one of Anish Kapoor's voids of pure black disorientates us, like the black holes of general relativity. And like the latter, it suggests that there are other ways of conceptualizing the impalpable fabric of reality.

Between observation and understanding the road can be long. Many great leaps forward in knowledge have been made through good use of brainpower alone, without *any* new observations. Scientific giants such as Copernicus and Einstein, for instance, obtained their momentous results based on observations that were already well known. In the case of Copernicus, they were known for more than a millennium. It is possible to discover the new even by starting from things that we already know, by leveraging details that do not add up, to find the gap toward the unknown. The ring that does not hold, the throw of the dice that does not add up (the gap is here?), the thread that, unraveled, can take us to the heart of a truth. The clue that may suggest to us how to rethink.

It is the capacity to change the organization of our thoughts that allows us to leap forward. Think about Copernicus. Before Copernicus, the world consisted of two great realms of things: the terrestrial (mountains, people, raindrops, etc.) and the heavenly (celestial bodies such as Sun and stars). Terrestrial things fall, celestial orbit. Terrestrial things are perishable, celestial are eternal. This is so reasonable that it takes reckless courage to propose a different way of organizing reality. Copernicus has it. His cosmos is divided in a different way. The Sun is in a class of its own. The planets are all in the same class, and the Earth is merely one of them. For all that it contains, for all its drops of rain, its people and its mountains, it is strangely in the same class as those points of light in the sky that are Venus and Mars. As for the Moon, well, it is in another class again, all by itself, poor little thing.* Everything revolves around the Sun, but the Moon revolves around the Earth. A subversion of reasonableness. A great one.

---

* It took Galileo's telescope, a century later, to end the moon's disturbing solitude, and find her sisters, the moons of Jupiter.

It is not easy to change the order of things—but this, at its very best, is what science does. Our conceptual structure is neither the definitive one nor the only one possible: it is rather the one that evolution has led us to cobble together in order to negotiate our daily needs. Often it does not work beyond that. Dividing everything into terrestrial and celestial objects works fine for everyday life, but not for understanding the cosmos and our position within it.

So how can we reconceptualize reality in order to cross the singularity that Einstein's equations predict in the future of a black hole? What is there, on the other side of the singularity? What is on the other side of Alice's looking glass?

What should we leave at home, and what should we take with us, to be light enough to pass through the mirror, beyond the end of time predicted by general relativity?

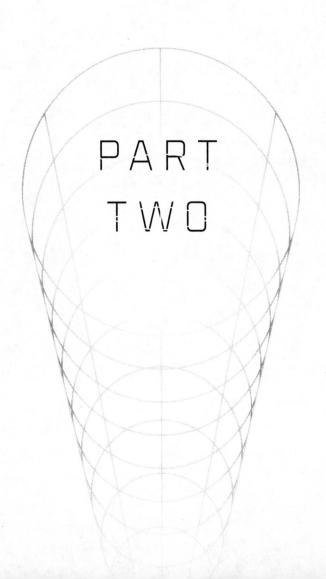

PART

TWO

We have arrived at that summer's day in my study, when after months of attempts, errors, false leads and discarded ideas, Hal suggested reversing time and linking two spacetimes with a quantum tunnel. What did he mean by this?

He meant to suggest what might be *beyond* the singularity.

The suggestion was based on the simplest of analogies. The formation of a black hole is a *fall*: a star that has finished burning *falls* upon itself, crushed under its own weight. An object that enters a black hole *falls*. Space itself, the long tube in the drawings above, is crushed by *falling* upon itself.

What happens to things when they fall? They reach the bottom, and then, quite often . . . they *bounce*. If I drop a basketball on the ground, it bounces and heads back upward.

How does the basketball move, after the bounce? Think about it for a moment. It moves as if a film of its fall were being played in reverse, backward in time. A ball that bounces is like a ball that falls, seen starting from the end of the process. As if the film of its fall were being run backward.

I have insisted on pointing out that the singularity of a black hole is not "in the center": rather, it is at the end of its fall. When the black hole arrives at the bottom of its fall, in the gray area in the last figure, could the star and the whole spacetime not simply bounce, and come back like a bouncing ball, as if time were reversed?

A few months before the discussion with Hal, I had studied, together with Francesca, my comrade, the possibility that a collapsed star could bounce at the end of its fall. We christened the bouncing star as a "Planck star,"[10] because the bounce is expected at the Planck scale.[11] Could the full black hole actually *bounce* around the star? What would we see if we were to imagine filming the full life of the black hole and then running the film backward?

We would see a white hole.

1

What, then, is a white hole?

Remember that we knew about black holes long before they were observed: they were solutions of Einstein's equations. A white hole is the same thing, a solution of Einstein's equations. So we know about white holes as well.

Actually, a white hole is not even *another* solution of Einstein's equations. It is the *same* solution that describes a black hole, reversed in time. The same solution with the sign of the time variable flipped. The same solution, as seen backward in time. A white hole is how a black hole would appear if we could film it and run the film in reverse.

Einstein's equations, like all equations of fundamental physics, do not specify a direction for time; they do not distinguish between past and future. If they allow a process to happen, they also allow that process to happen backward in time.*

So if a black hole rebounds and retraces its previous route in time, like a basketball bouncing, this means that it is transformed into a white hole.

Here is a drawing of how the internal space of the black hole can continue.

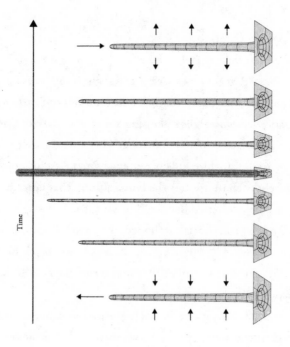

---

* If this statement puzzles you, please hold on until the third part of this book.

When it enters the quantum zone (gray), the tube stops lengthening and narrowing, and bounces. It goes back, now shortening and widening back.

In a black hole you can enter, but not leave. In a white hole, on the contrary, you can exit, but not enter. (A film of things going into a hole shows things coming out of the hole, if run backward.) So anything that enters a black hole can cross the gray zone, pass into the white hole, and then come out again.

Quite straightforward, no?

IS IT REALLY POSSIBLE FOR THIS TO HAPPEN? TO PASS from a black hole to a white hole, space and time must cross the gray zone in the figure above. This violates Einstein's equations. It may be only for a brief moment—the moment of the rebound—but it is still a violation. We do expect Einstein's equations to be violated: quantum effects must come into play. But do these effects allow for a rebound? The quantum physics of atoms, of electrons, of light and of lasers are well known to physicists. But we are dealing here with the quantum physics of space and time.

And this is precisely the reason black holes and white holes interest me so much. I have spent my life attempting to understand precisely these quantum aspects of space and time, trying to unravel the conceptual structure needed to navigate when space and time are quantum. It is my passion. *I know the signs of the ancient flame!** Down in the bottom of the black hole, I see her gleaming.

For the most part, my work in theoretical physics has been to participate in the construction of a mathematical structure capable of dealing with quantum spacetime. The mathematical structure we have constructed is called *loop quantum gravity*. To understand what happens in the regions of a black hole where the quantum aspects of space and time dominate, and where the continuous space and time of our experience no longer functions, this theory is needed. Here we shall see if it works. *Here your worth will be tested.*†

---

* *Purgatory*, XXX. These are Dante's words as, halfway through his voyage, he perceives the presence of his great love, Beatrice.

† *Inferno*, XX. Dante addresses these words to himself, perhaps as an encouragement, a hope or a warning, early in his poem.

2

What does "quantum behavior" mean?*

The simplest quantum property is granularity. At a small scale, all processes manifest themselves in a granular way. Light observed at low intensity manifests itself in *grains* of light: photons.

Applied to space, this basic idea implies that there should be *elementary grains of space*, of finite size. Quanta of space. That is, there are no arbitrarily small things; there is a lower limit to divisibility. Space is a physical entity, and like any other entity, it is granular. Einstein's theory and the mathematics of quantum theory combine to imply this fact.[12]

The mathematics required to obtain this result was developed years ago by Roger Penrose, the great English relativist

---

* I have written a book, *Helgoland*, seeking to answer this general question as best as I see it possible.

awarded the Nobel Prize while I was writing the first draft
of this book. This math as well was born of a simple analogy:
a net. A net is a set of nodes connected by links. The nodes
denote the elementary grains of space. They are the "quanta
of space," just as photons are the quanta of light. But there
is a fundamental difference: whereas photons move *within*
space, the quanta of space are the grains that weave the net
that *is itself* space.

The links that join the nodes indicate which quanta are
adjacent, thus defining a connected, or "spatial," structure.
Roger Penrose has called these structures "spin networks."
The term "spin" comes from the mathematics of the sym-
metries of space, where rotation, or spin, plays an important
role.[13] Here is a sketch of the relation between a network and
the chunks of space it describes.

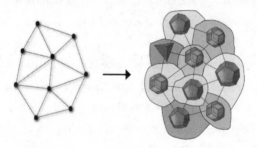

In 1958, Penrose, the Englishman, met Finkelstein, the
American, who understood how horizons work and who wrote
about the engravings of Dürer. Finkelstein had traveled to

London to give a lecture on the horizons of black holes, which he had just unraveled. Penrose had recently finished his studies at Oxford and came to London to listen. After the lecture, the two young men had a long discussion. Penrose had already begun to develop the rudiments of the mathematics of spin networks, and during the conversation he introduced this math to Finkelstein.

They both emerged changed by the encounter. Penrose got passionate about black holes. The passion, ignited by Finkelstein's lecture, led him in the following years to demonstrate that the formation of black holes is inevitable—as a result of which, sixty years later, while I am writing this book, he will receive a Nobel Prize. Finkelstein, for his part, got passionate about the discrete structure of space that Penrose had begun to investigate by inventing spin networks. He will search for a long time for a quantum description of a spacetime made up of elementary quanta. In a peculiar interaction, the two adventurers, explorers in the land of ideas, swapped interests.[14]

I was two years old at the time of this conversation. Thirty-five years later, Lee Smolin and I stumbled upon Penrose's mathematics of spin networks and the granular space they describe, by applying quantum theory techniques to general relativity—thus combining the two research worlds that Penrose and Finkelstein had swapped three decades earlier.

At the time (it was 1994) Lee came to visit me often in

Verona. (Not on my account alone, I later realized; he was fascinated by a beautiful Veronese friend of mine.) Calculating the properties of the elementary quanta of space, we realized that we were rediscovering Penrose's spin networks. Lee flew from Verona to Oxford to ask Roger to explain to him the details of his math. Ever since, Roger Penrose has been like a wonderful older brother to us. But let's get back to black holes.

If space is granular, the inside of a black hole cannot be squashed any further than the size of the individual grains. There is a limit to compression and distortion. The contraction that squeezes the tube of the inside of a black hole must therefore stop *before* the singularity. What happens next?

3

The second main feature of quantum phenomena is that the properties of things are not always definite. A particle does not always have a position. Sometimes it can be nowhere, disembodied like a wave, before materializing somewhere else: it can *leap*.

One consequence of this leapy aspect of reality is the phenomenon known as the "tunnel effect." This is the ability that things have to cross barriers that would be insurmountable if it weren't for the quantum. Imagine throwing a marble at a wall. Classical physics (and common sense) tells us that it cannot pass through the wall. In reality, the marble has a minuscule probability of passing through it to the other side. This is the tunnel effect. It is called this because it is as if the marble could find a "tunnel" (an imaginary one) that allows it to cross through any barrier.

This was Hal's first idea: the inside of the black hole is able to cross the zone forbidden by Einstein's equations—the gray zone in the figures above—and jump, by tunnel effect, "to the other side."

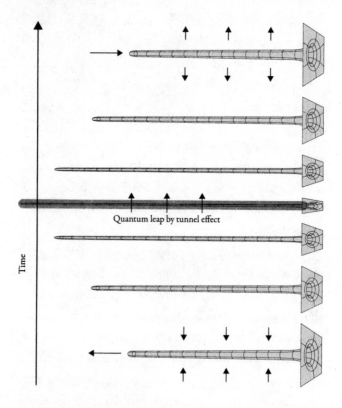

Quantum leap by tunnel effect

Time

The quantum properties of space and time allow the inside of the black hole to "leap" beyond the singularity, when classical equations would have time stop.

Quantum leaps are well known in physics. Since the early days of the quantum saga, Niels Bohr realized that atoms emit light when electrons "jump" from a larger orbit to a smaller one. But here the quantum leap is far more radical than a particle hopping around from one location to another: here it is spacetime itself that jumps.

The leap of spacetime is not a phenomenon that takes place in space and in time. It is a phenomenon that is neither spatial nor temporal: it is an instantaneous quantum transition from one configuration of space to another. Quantum transitions of this kind—leaps from one configuration of space to another—are precisely what is described by loop quantum gravity.

The equations of ordinary quantum mechanics give the probabilities that jumps from one configuration to another will occur for a physical system that is *in space*. The equations of loop quantum gravity give the probabilities of leaps from one configuration *of space* to another configuration *of space*.

Crossing the zone where Einstein's theory predicts the end of time, time and space do not exist.

So here we are. Here the quantum properties of time and space flare. *I know the signs of that ancient flame!* We can cross what Einstein's theory defined as the edge of reality. We can pass to the other side. The equations of loop quantum gravity allow us to calculate the probability of this happening.

THIS IS THE KEYSTONE. IT IS THE KEYSTONE BOTH OF the scientific problem—the fate of black holes—and of this book. The leap beyond the end of time predicted by general relativity can happen: it is predicted by quantum theory. Like all quantum "leaps," it is a break, a rupture of continuity. A momentary fracturing of the flow. And yet it is described by equations that we have. The equations of quantum gravity describe a world more complex than a simple spatiotemporal continuum.

Having ascended into the skies of Purgatory and arrived at the edge of the known universe, Dante loses Virgil. But at the same moment, filled with emotion—*I know the signs of the ancient flame!*—he sees Beatrice.

A short while later, with a dazzling play of looks between Dante, Beatrice and the sun, Dante ascends from Earth to Heaven, from the known part of the cosmos* to the other. Beatrice looks at the sun, Dante looks at Beatrice's eyes.

> *Beatrice upon the eternal wheels had fixed*
> *Her eyes; and I on hers fixed mine . . .*

---

* Dante's Hell is a vast funnel carved inside the Earth, and his Purgatory is a mountain in the Southern Hemisphere. The quotations that follow are from Canto I of *Paradise*.

Following Beatrice's gaze, Dante himself stares into the sun. He is flooded with light . . . *It seemed then that so much of the sky was lit.* There is nothing else, in the leap, but a lake of light . . . *And suddenly it seemed that a sun was added to the sun* . . . he is lost again in the eyes of Beatrice . . . and then . . .

Much that our powers here cannot sustain,
is permitted there . . .

. . . we fly to the other side of space and of time.

4

et's redraw the image of the transition from black hole to
white hole, with a few more details added.

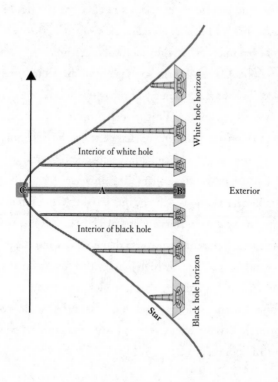

I have added the trajectory of the star that forms the black hole around itself, bounces and in the end comes out from the white hole. The star always remains at the bottom of the long funnel. On the right of the figure I have also added the *exterior* of the hole.

The gray area is the zone of quantum transition. Everything outside this gray area is in accord with Einstein's theory.

I have divided the quantum transition zone into three parts and labeled them A, B and C (see the last figure), because that is what they are called, not very imaginatively (and not even in order), in the technical literature.

Zone A is the internal passage from the geometry of the black hole to the geometry of the white hole. Zone C is the rebound of the star. This is very similar to what, according to loop quantum gravity, happened at the Big Bang. The Big Bang may have been a large cosmic rebound (or "Big Bounce"), in which a contracting universe reaches the maximum density allowed by the quanta, then rebounds and begins to expand. In the case of a black hole, it is just the star rather than the entire universe that bounces, but the physics is similar: at extremely high density the quanta are discrete and generate a pressure that impedes any further degree of compression, causing the rebound. In both cases it is quantum gravity that causes the pressure that turns a collapse into a bounce.

At the point of maximum compression, the extremely dense star is the Planck star.[15] "Planck star" is also the name given to the entire phenomenon: the star that collapses into a black hole, the rebound, the white hole, until the point when everything comes out again.

The area that is most difficult to treat mathematically is zone B, the quantum leap of the horizon from black to white. The calculations for this transition are currently in progress. They are based on a version of loop theory called "covariant," or more colorfully, "spinfoam."

i'm rereading these lines for the umpteenth time. i am in verona, in the square named after the poet. in front of me, his austere statue. i am sitting on the steps of the loggia di fra giocondo. here i first saw my first love.

this is the city of my birth, to which, whenever i can, i return from my (happy) exile in various parts of the world. for dante it was a place of painful exile, where he learned *how the bread of others tastes of salt* (bread is salted in verona, unlike in florence, his birthplace), and *how hard the way is, ascending and descending stairs of others*. in front of me is the palazzo della ragione, with its very long staircase. the palazzo would have been here seven hundred years ago. perhaps not the staircase. dante would certainly have often been in this exact spot where i am now. he wrote the *paradiso* here. he must have sat in this piazza, on these steps, with a manuscript . . .

nearby is the little church of sant'elena, with an enchanting cloister where i, as a kid, used to go stealing kisses with girls (until caught by an austere priest). in the cloister there is what is perhaps the oldest library in the world (i have held parchment manuscripts from the third century and codices from the fifth in my hands). here dante gave a lecture, *a question of the water and of the land*, in which he discusses why, if the natural place of land is below water, there is nevertheless land that is prominent above sea level. it is a good question. some say he gave the lecture in pursuit of a teaching position at the "studio," the school of verona that was developing into a university. who knows if that is true. for sure he did not get a job. probably he was judged not sufficiently qualified—or not of the right, clubbable sort. he, who sang the universe as he did . . .

I digress. Let's return to the transition between black holes and white holes. We are now clear of the star, clear of its interior and clear of the horizon.

But there is more. We are missing the most important step: if this is what happens *inside* the hole, what happens *outside*? How can *the outside of a black hole* transform into *the outside of a white hole*, if outside we do not expect anything that is quantum?

In order to answer this question, and to understand the second insight that Hal had on that day, we need a better understanding of what a white hole is.

Be prepared to be surprised.

5

How does the *exterior* of a white hole differ from the exterior of a black hole? If we are outside, how can we distinguish a black hole from a white one?

The surprising answer is that we can't. Unless something goes in or out of the horizon, from the outside a white hole is indistinguishable from a black hole.

A black hole attracts, like every mass—the same goes for a white one. Around a black hole there may be planets in orbit—the same goes for a white hole. And so on. You can fall toward a black hole, and you can fall toward a white hole.

This is confusing. A white hole is like a black hole rebounded. One may be tempted to expect that, reversing time, gravitational attraction becomes repulsion. But it is not so. Gravitational attraction does not become repulsion by reversing the direction of time:[16] in a film of the solar system projected backward, the Sun is still *attracting* the planets; in a film of a stone that is thrown upward and then falls, we still see the stone flying upward and then downward, *attracted* by the Earth, even if the film is projected in reverse. The outside of the black hole does not change if the direction of time is reversed: seen from the outside, a black hole and a white hole behave in the very same way. They are both masses that attract through the force of gravity.

HOW CAN THIS BE? THEY SEEM LIKE SUCH DIFFERENT objects: you can only enter a black hole, and you can only exit a white hole. How is it possible that, despite this, they are indistinguishable? It seems like something of a paradox.

It is not. And it is here that the extraordinary architectural magic of general relativity shines through. It is a delicate, beautiful point. I explain this in detail in the rest of this section. It is subtle and intricate, and most people get lost here. No matter if you do; it is not important for what follows. You can skip the rest of this section without compromising your understanding of the whole story. But if you do manage to follow, you will be amazed by what the relativity of time encompasses.

Here we go. A stone can exit from a white hole. Hence, I can see a stone moving freely away from a white hole. Can I see a stone moving freely away from a black hole? At first sight this seems hard: how can a stone move freely away from a black hole if *nothing can exit* from it? And yet, it is possible. Here is how. If someone throws a stone with great force away from the collapsing star *a moment before crossing the horizon*, the stone will fly away. But seen from afar, the first part of the flight will take place extremely slowly, since everything is in extreme slow motion here, seen from afar. The stone will end up moving away from the black hole for good only after a very long (external) time. Hence, we can see it emerging *at any later time*. Thus, it is not impossible for us to see a stone freely moving away from a black hole, just as it is possible to see one moving away from a white hole.[17]

The argument also works the other way around. A stone falling toward a white hole cannot cross the horizon, because nothing can enter the horizon of a white hole. This would seem to indicate that from the outside we can easily distinguish between a black hole and a white hole: it suffices to observe a falling stone. If we see it entering the hole, the hole must be black. But remember, you never see a stone *enter* the horizon of a black hole! Since light takes longer and longer to travel away, we can see it *approaching* the horizon gradually, without ever seeing it enter. And we see the very same of a stone that falls toward a white hole! We see it approaching the horizon gradually, without entering it.

So what happens *to the stone* that is falling toward a white hole? It eventually collides with the material that is coming out of the white hole: this happens in a time that is very short for the stone itself, which reaches the outgoing matter before reaching the horizon; but this time is extremely long when viewed from the outside (once more: time slows in the vicinity of the horizon!).

This is the magic of the elasticity of time. The black and the white horizons are themselves different, but from the outside they look precisely the same. The horizons and the insides distinguish white from black, future from past; the outside does not.

Recall the title of the 1958 work in which David Finkelstein showed what happens at the horizon: "Past-Future Asymmetry of the Gravitational Field of a Point Particle." The title underlines the central insight: the geometry of the *outside* of a black hole does not change with the direction of time, but this symmetry is broken at the horizon. The horizon is not invariant with the reversal of time—the exterior is. It is for this reason that the same exterior is compatible with both a black hole and a white hole, despite their horizons being so different.

All of this seems almost incredible, but this is how nature works. Despite the complete difference between what happens *inside* them, time's sleight of hand on the horizons allows a white hole and a black hole to be the same thing *outside*.

And *this* was the crucial observation that Hal made on that day.

—◦◦—

WHY? BECAUSE IT MAKES IT PLAUSIBLE THAT WHAT
happens inside the black hole is exactly what is illustrated in
the figure on page 79. The magic is that *inside* the horizon the
space evolves as in the drawing, whereas outside . . . nothing
at all happens!

The quantum tunnel occurs only in areas of great spatiotemporal distortion, in a way compatible with the fact that,
on the outside, where there is nothing quantum, everything
continues to respect general relativity as it should.

The solution of Einstein's equations that black holes represent and the one represented by white holes can therefore
be *glued* together on the outside, without violating Einstein's
equations. This is the "pasting" Hal suggested. The only violation occurs where we expect it, where the distortions are
so strong that they produce quantum effects.

Bingo! We have found a plausible scenario for what happens beyond the stopping of time. Beyond the singularity,
there is the solution with time reversed: the inside of a white
hole. Outside, nothing happens. The black horizon, like Gandalf, has magically turned white.

—◦◦—

I REMEMBER THE EMOTIONS OF THAT DAY, WHEN THE
scenario intuited by Hal began to come into focus. The pieces

of the puzzle were known: the tunnel effect, the solution of Einstein's equations with white holes, the solution with black holes, the existence of a lower limit to the divisibility of space, the strange behaviors of white and black holes, the scandalous temporal differences between what happens on the horizon and from afar, the intuition that things that fall rebound and that Planck stars can do so, too.[18] The pieces of the puzzle fit together.

As in all such scientific puzzles, certain pieces fail to fit and are discarded. What happens *precisely* to space and to time at the moment of the bounce? Quantum theory says that what happens *during* the leap does not exist. It has neither shape, dimension, nor any other properties for that matter.

We can get a rough idea of what happens if we imagine the crushing of the tube gently slowing, then rebounding from its direction of travel and beginning to expand. But the reality is that, in this transition, space and time dissolve in a cloud of probability, after which they resume their structure. The piece of the puzzle to jettison is the idea that events in nature can always be imagined as if they were taking place in space and in time.

THERE WERE MANY QUESTIONS THAT REMAINED UN-answered that evening. We had to do the calculations. Good analogies are nice, but to check that they are not illusory, you need syllogisms. We needed to get down the equations to describe the geometry of our spacetime—to demonstrate that they are consistent with Einstein's equations, up to the transition. Set up the calculation of the probability for the quantum leap.

We did this in the days that followed. Exciting and fun. Cutting and stitching, ensuring that pieces fit. The problem was that each portion of spacetime was described using a perspective that did not include the others, which was precisely the issue that had confused Einstein and everybody else from the very beginning, and that Finkelstein had clarified. Techniques for addressing this are now familiar;[19] we used them. everything worked. we wrote an article with the results and published it.[20] slowly, the idea began making its way.

the hypothesis that a black hole can transform into white was now in the hands of whoever wanted to help it grow.

YES, WE WERE HAPPY THAT NIGHT   FEW THINGS ARE as good as the rarefied lightness of this feeling   having had a good idea that might turn out to be correct   a calculation that finally adds up   an insight into how something might work   something we did not understand before   a subtle happiness, subcutaneous but pervasive, as if suddenly feeling right with the world.

perhaps it is just the feeling of having done a job well. the same i get after having fixed the garden gate. doing science is usually a succession of disappointments, things that don't work, wrong ideas, experiments that fail, calculations that do not add up—occasionally punctuated by moments of joy. perhaps there was something else. the joy of a step that satisfies a little our desire to understand, to "go and see" . . . yes, we were very happy that evening, hal and i.

and yet, for all this, we were for sure far from being convinced we were in possession of a truth. science is replete with delusions—might this be one, too?

years have since passed. the idea of a transition from black to white has developed and been studied by many, in various forms. we are looking for evidence in the sky. and yet, even today, i am far from convinced that we have the truth in our pocket.

scientists have a difficult relationship with their own ideas: perhaps no one is completely honest, even with oneself, about how much one believes . . . you need to be politically correct, reasonable, admit that you could be wrong. but at heart there is a mad desire to scream, "but i'm sure that this is how things are!" one falls in love with his own ideas, is convinced by them . . . defends them tooth and nail. after all, scientific reputation, which we cling to like children to candy, depends on this. and yet, and yet . . . at the same time, down at the bottom of our hearts, doubt is never quelled . . . the fear that we are wrong, we are deluded . . . *bitter-sweet is science.*

Paul Dirac, the most rational, impassive, cerebral, autistic of scientists, remarked in a lecture that the reason it is rare for a good scientist who has obtained an important result to take the next step is that he is the first to have doubts about his own results. Dirac tells how, when he found the equation that today bears his name—one of the most celebrated equations in modern physics, describing how electrons move— he immediately published the calculation that showed that the equation gave the right prediction for atomic spectra *in the first approximation*. He did not have the courage to make the calculation at a better approximation, fearing that he would get it wrong and show everyone that his equation was wrong.

will our idea hold? i ask myself, walking beneath the huge trees in the wood behind the house. it seems obvious to me that the idea must be right. really, what else can *reasonably* happen, when everything is taken into account? i have turned it over in my mind in every possible way, cannot see where it could be wrong. other times i smile to myself, remembering how many misconceptions there have been and there are, that seemed and seem so right to those working on them . . .

. . . doubts or no doubts, hopes or fears, we were happy that evening. a good day. a step forward, we are not sure where toward. we live also for this.

PART

THREE

t will not have escaped the insightful reader that the key to Hal's idea was *time*. A white hole is a black hole *with time reversed*.

But is such reversal of time really possible? Most phenomena occur in one direction only: reversing them in time is *not* possible. A dropped egg does not bounce. Past and future *are* different.

The reconstruction of the life of a black hole I have given so far is too simple, for it neglects everything that distinguishes the past from the future, and certainly there are many phenomena that do so. To complete the picture, we need to take into account phenomena that *cannot* be reversed in time—the "irreversible" aspects in the life of black holes. This leads us—once again—toward questions about the nature of time.

What makes past and future different? Why are they

different at all? Why do we remember the past and not the future? Why is it possible to decide what we shall do tomorrow, but not what we did yesterday? I have been working closely with these questions over the last few years.

This last part of the booklet is about the *irreversible* aspects in the life of a black hole. I start with an amusing ongoing controversy that has scientists squabbling. Then I want to tell you about a number of things I think I have recently understood about the direction of time—and that seem quite beautiful to me.

1

I n 1974, Stephen Hawking made an unexpected theoretical discovery: black holes must emit heat.[21] This, too, is a quantum tunnel effect, but a simpler one than the bounce of a Planck star: photons trapped inside the horizon escape thanks to the pass that quantum physics provides to everything. They "tunnel" beneath the horizon.

So black holes emit heat, like a stove, and Hawking computed their temperature. Radiated heat carries away energy. As it loses energy, the black hole gradually loses mass (mass is energy), becoming ever lighter and smaller. Its horizon shrinks. In the jargon we say that the black hole "evaporates."

Heat emission is the most characteristic of the *irreversible* processes: the processes that occur in one time direction and cannot be reversed. A stove emits heat and warms a cold room. Have you ever seen the walls of a cold room emit heat and heat up a warm stove? When heat is produced, the process is irreversible. In fact, whenever the process is irreversible, heat is produced (or something analogous).[22] Heat is the mark of irreversibility. Heat distinguishes past from future.*

There is therefore at least one clearly irreversible aspect to the life of a black hole: the gradual shrinking of its horizon.[23]

But, careful: the shrinking of the horizon does not mean that the *interior* of the black hole becomes smaller. The interior largely remains what it is, and the interior volume keeps growing. It is only the horizon that shrinks. This is a subtle point that confuses many. Hawking radiation is a phenomenon that regards mainly the horizon, not the deep interior of the hole. Therefore, a very old black hole turns out to have a peculiar geometry: an enormous interior (that continues to grow) and a minuscule (because it has evaporated) horizon that encloses it. An old black hole is like a glass bottle in the hands of a skillful Murano glassblower who succeeds in making the volume of the bottle increase as its neck becomes narrower.

At the moment of the leap from black to white, a black hole can therefore have an extremely small horizon and a vast interior. A tiny shell containing vast spaces, as in a fable.

---

* I discuss this point extensively in my book *The Order of Time*.

⬍

IN FABLES, WE COME ACROSS SMALL HUTS THAT, WHEN entered, turn out to contain hundreds of vast rooms. This seems impossible, the stuff of fairy tales. But it is not so. A vast space enclosed in a small sphere is concretely possible.

If this seems bizarre to us, it is only because we became habituated to the idea that the geometry of space is simple: it is the one we studied at school, the geometry of Euclid. But it is not so in the real world. The geometry of space is distorted by gravity. The distortion permits a gigantic volume to be enclosed within a tiny sphere. The gravity of a Planck star generates such a huge distortion.

An ant that has always lived on a large, flat plaza will be amazed when it discovers that through a small hole it has access to a large underground garage. Same for us with a black hole. What the amazement teaches is that we should not have blind confidence in habitual ideas: the world is stranger and more varied than we imagine.

The existence of large volumes within small horizons has also generated confusion in the world of science. The scientific community has split and is quarreling about the topic. In the rest of this section, I tell you about this dispute. It is more technical than the rest—skip it if you like—but it is a picture of a lively, ongoing scientific debate.

The disagreement concerns how much *information* you

can cram into an entity with a large volume but a small surface. One part of the scientific community is convinced that a black hole with a small horizon can contain only a *small* amount of information. Another disagrees.

What does it mean to "contain information"?

More or less this: Are there more things in a box containing five large and heavy balls, or in a box that contains twenty small marbles? The answer depends on what you mean by "more things." The five balls are bigger and weigh more, so the first box contains more matter, more substance, more energy, more stuff. In this sense there are "more things" in the box of balls.

But the number of marbles is greater than the number of balls. In this sense, there are "more things," more details, in the box of marbles. If we wanted to send signals, by giving a single color to each marble or each ball, we could send more signals, more colors, more information, with the marbles, because there are more of them. More precisely: it takes more *information* to describe the marbles than it does to describe the balls, because there are more of them. In technical terms, the box of balls contains more *energy*, whereas the box of marbles contains more *information*.

An old black hole, considerably evaporated, has little energy, because the energy has been carried away via the Hawking radiation. Can it still contain *much* information, after much of its energy is gone? Here is the brawl.

Some of my colleagues convinced themselves that it is not possible to cram a lot of information beneath a small surface. That is, they became convinced that when most energy has gone and the horizon has become minuscule, only little information can remain inside.

Another part of the scientific community (to which I belong) is convinced of the contrary. The information in a black hole—even a greatly evaporated one—can still be large. Each side is convinced that the other has gone astray.

Disagreements of this kind are common in the history of science; one may say that they are the salt of the discipline. They can last long. Scientists split, quarrel, scream, wrangle, scuffle, jump at each other's throats. Then, gradually, clarity emerges. Some end up being right, others end up being wrong.

At the end of the nineteenth century, for instance, the world of physics was divided into two fierce factions. One of these followed Mach in thinking that atoms were just convenient mathematical fictions; the other followed Boltzmann in believing that atoms exist for real. The arguments were ferocious. Ernst Mach was a towering figure, but it was Boltzmann who turned out to be right. Today, we even see atoms through a microscope.

I think that my colleagues who are convinced that a small horizon can contain only a small amount of information have made a serious mistake, even if at first sight their arguments seem convincing. Let's look at these.

The first argument is that it is possible to compute *how many* elementary components (how many molecules, for example) form an object, starting from the relation between its energy and its temperature.[24] We know the energy of a black hole (it is its mass) and its temperature (computed by Hawking), so we can do the math. The result indicates that the smaller the horizon, the fewer its elementary components.

The second argument is that there are explicit calculations that allow us to count these elementary components directly, using *both* of the most studied theories of quantum gravity—string theory and loop theory. The two archrival theories completed this computation within months of each other in 1996.[25] For both, the number of elementary components becomes small when the horizon is small.[26]

These seem like strong arguments. On the basis of these arguments, many physicists have accepted a "dogma" (they call it so themselves):[27] the number of elementary components contained in a small surface is necessarily small. Within a small horizon there can only be little information. If the evidence for this "dogma" is so strong, where does the error lie?

It lies in the fact that both arguments refer *only* to the components of the black hole that can be detected from the outside, as long as the black hole remains what it is. And these are only the components *residing on the horizon*. Both arguments, in other words, ignore that there can be components in the large interior volume. These arguments are formulated from the perspective of someone who remains far from the black hole, does not see the inside, and assumes that the black hole will remain as it is forever. If the black hole stays this way forever—remember—those who are far from it will see only what is outside or what is right on the horizon. It is as if for them the interior does not exist. *For them*.

But the interior does exist! And not only for those (like us) who dare to enter, but also for those who simply have the patience to wait for the black horizon to become white, allowing what was trapped inside to come out. In other words, to imagine that the calculations of the number of components of a black hole given by string theory or loop theory are *complete* is to have failed to take on board Finkelstein's 1958 article. The description of a black hole from the outside is incomplete.

The loop quantum gravity calculation is revealing: the number of components is precisely computed by counting the number of quanta of space *on* the horizon. But the string theory calculation, on close inspection, does the same: it assumes that the black hole is stationary, and is based on what is seen from afar. It neglects, by hypothesis, what is inside and what will be seen from afar after the hole has finished evaporating—when it is no longer stationary. (The inside of the black hole, remember, is not at all stationary: it changes. The long tube elongates and narrows.)

I think that certain of my colleagues err out of impatience

(they want everything resolved before the end of evaporation, where quantum gravity becomes inevitable) and because they forget to take into account what is beyond that which can be immediately seen—two mistakes we all frequently make in life.

Adherents to the dogma find themselves with a problem. They call it "the black hole information paradox." They are convinced that inside an evaporated black hole there is no longer any information. Now, everything that falls into a black hole carries information. So a large amount of information can enter the hole. Information cannot vanish. Where does it go?

To solve the paradox, the devotees of the dogma imagine that information escapes the hole in mysterious and baroque ways, perhaps in the folds of the Hawking radiation, like Ulysses and his companions escaping from the cave of the cyclops by hiding beneath sheep. Or they speculate that the interior of a black hole is connected to the outside by hypothetical invisible canals . . . Basically, they are clutching at straws—looking, like all dogmatists in difficulty, for abstruse ways of saving the dogma.

But the information that enters the horizon does not escape by some arcane, magical means. It simply comes out after the horizon has been transformed from a black horizon into a white horizon.

In his final years, Stephen Hawking used to remark that there is no need to be afraid of the black holes of life: sooner or later, there will be a way out of them. There is—via the child white hole.

WHERE THERE ARE DISAGREEMENTS, STILL, THERE ARE also doubts. what if the others were right instead of us? what is to be done? read, attempt to understand their reasoning, question ourselves . . .

but then, if in the end it still seems to us that they are mistaken, we must have the courage to listen to the voice of the master: *let the people talk: stand firm as a tower that does not shake, even though buffeted up there by the winds.**

after all, this, in the end, is what it means to do science. the aim is not to convince those around you. the aim is to truly understand. clarity will find its way, following its own course, in its own time. it takes infinite humility to trust in yourself. but also infinite arrogance, to have the strength to go *along the solitary plain.*† as all those who paved the way for us have done.

---

* *Purgatory*, V. This is Virgil telling Dante to keep to his way, unafraid of what the people murmur.

† *Purgatory*, I. Dante, in a moment of solitude.

I HAVE TWO READERS IN MIND WHEN I WRITE. ONE knows nothing about physics: I try to communicate to him the charm of research. The other knows everything, and I try to offer her new perspectives on what she already knows. For both, I aim at the core of the matter. I remove from my writing anything I can. I imagine that those who know nothing of physics would find details useless and burdensome. The experts, on the other hand, know the details already; they are not interested in hearing them repeated. They want a novel perspective.

In this way, however, I displease and sometimes even annoy intermediate types of readers—those who are acquainted with the field but perhaps have not yet immersed themselves in it completely. Physics students, for example. The worst reviews I get come from them. I understand. It grates to see details skipped that you have painstakingly studied, and to find things presented in a way that is different from the sacred (text)books, for which I can only apologize.

But there is another reason I sometimes annoy my young colleagues: I don't use the jargon, the parlance of the trade. Imagine how shocked a sailor would be if instead of "Ease the jib!" he heard you shout, "Let go a little the rope attached to the small sail!" For the uninitiated, however, "Let go a little the rope attached to the small sail!" will surely be more comprehensible than "Ease the jib!" Reading the last few pages, students who have recently learned these things will be tearing their hair out and wondering why on earth Rovelli doesn't just call things by their proper names. I'll try to remedy that: in an endnote I've provided a translation of the previous few pages into the jargon of physics. It says the same, in technical terms. My initiated readers will feel a little more at home and find the argument a little more precise.[28]

2

Let's move on, leave the controversy around the "information paradox" (which is not a paradox) and get back to where we were. Hawking radiation is irreversible, like a cooling hot stove. Therefore, the life of a black hole cannot be reversible. The rebound cannot be complete.

Think again about the ball that bounces on the ground. I wrote that the upward bounce mirrors the movement of its fall, seen reversed in time. But this is not exactly true. Air resistance slows the fall, and it slows the rise as well. The bounce on the ground is not perfectly elastic: it leaves an imprint. These are irreversible phenomena. They cause the ball to dissipate energy in the form of heat. As a consequence, the rise of the ball after the bounce is not exactly the same as the fall: the ball does not return to the same height from which it fell. The bounce of the ball, in other words, is reversible only at a first approximation. Looking more closely, we notice irreversible phenomena that render the process not really symmetrical in time. Past and future are different.

The same is true for a Planck star. A black hole loses energy when it emits Hawking radiation. It becomes smaller—and when the star rebounds into a white hole, it does not get back to being as big as the black hole was to start with. It stays small. The white hole that forms is smaller than its parent.

Hawking radiation can shrink the horizon until it is very small indeed. At this point, the spatiotemporal distortion around the horizon is very great. We are in a full quantum regime, and the probability of the leap from black to white becomes very large. The leap happens.[29] The white hole does not have the energy to regrow; it remains very small, emitting very weak radiation for a very long time,[30] before disappearing altogether.

The paths of energy and information throughout the course of a Planck star's life are thus very different. Almost all of the star's initial energy is lost through Hawking radiation. The way in which *the star* loses its energy is curious, and genuinely quantum: Hawking radiation has a component of negative energy (yes, energy can also be negative in the quantum world!) that goes inside the black hole. This eats away at the mass of the black hole and ends up on the star, annihilating its energy. Very little residual energy reaches the white hole horizon. This is the flow of the major part of the energy:

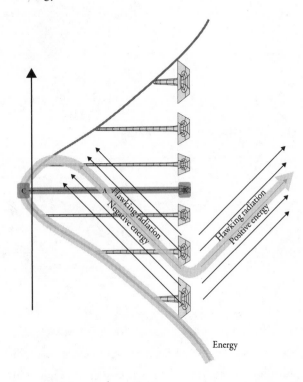

Energy

Information entering the horizon, on the other hand, remains trapped until after the quantum leap. The leap frees it, *to return to the world of light.*

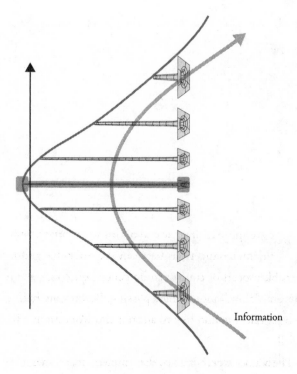

Information

It takes a long time to release a lot of low-energy information from a very small horizon (think of an enormous
number of very small marbles needing to get through a tight
opening). The information needs a long period to exit. The
white hole must live for a long time.

When all of the information and residual energy inside
has finally left, the long, happy life of the rebound of a Planck
star is over, and the white hole horizon dissipates.

3

We are approaching the end of this story. But the subtle interconnections between the reversible and irreversible aspects of time open up general questions about temporal "flow," about time's passing. Before concluding my brief tale, I would like to address those questions a bit as well.

The bounce is permitted by the symmetry that comes from the reversal of time. But time maintains its direction nevertheless: the precise moment of the bounce is symmetrical under time reversal, but the whole process is not. The tremendous temporal distortions effected by black holes and white holes play havoc with our perception of time, but they do not affect its orientation, the eternal current. The past remains different from the future. How so?

Physics tells us something very strange indeed about the

direction of time, and no doubt many questions would have occurred to the astute reader when I wrote above that the fundamental equations of physics do not distinguish between past and future. No direction of time comes from these equations. And yet, I then went on to speak about phenomena oriented in time. Where does the direction of time come from, if it is not inscribed in the fundamental grammar of the world?

It comes from the fact that we live in just one of the many solutions of the fundamental equations, and *in this solution*, the past appears special to us. The difference between past and future, that is, is a bit like the difference between two geographical directions for someone living in the mountains: in one direction—say north—the ground goes up; in the other—say south—it goes down. But this is not because north and south are intrinsically connected to up and down. Rather, it is only because things happen to be arranged in this way there. On the Italian side of Monte Bianco, "upward" is north, whereas on the French side, it is south. The irresistible flow of time is similarly a reflection of the way in which things happen to be arranged around us.

The same goes for a Planck star. The difference between past and future does not come from a fundamental asymmetry intrinsic to time. It comes from the fact that the past happened to be particular, like the top of the mountain in the geography example. Think about it: in the future, Hawking radiation fills the sky with energy, dissipating it everywhere. In the past, instead, energy was concentrated in the collapsing star. The past was particular, because the energy was concentrated, instead of being dissipated as, in general, it is naturally found to be in the future. The direction of the past is special, just as in any mountain village the direction toward the mountaintop is "special."

I have written at length about this basic equivalence between past and future in my book *The Order of Time*. But it is not easy to digest this idea: it goes against our most basic intuitions. Is it really possible that *all* differences between past and future come down to being just a consequence of how things happened to be arranged in the past?

Our intuition suggests the opposite: it appears to tell us that the past is radically different from the future. The very nature of reality, in our intuition, is a flow in a directional time. The past is determined, whereas the future is open. Is it possible that our sense of things can be so very wrong? If it is wrong, why do we intuitively get things so deeply wrong?

I asked myself these questions frequently over years of feverish work on white holes and their temporal distortions, zigzagging between the reversible and irreversible aspects of horizons.

There are two phenomena that we know intimately that radically distinguish past from future. They appear to be so basic and banal as to make it impossible to entertain the idea that time by itself has no direction. They are two very vivid asymmetries between past and future, which seem irreducible. The first is that we *know* the past but not the future. So the past appears fixed, determined. The second is that we can *choose* the future, but we cannot choose the past. The future appears to be open, indeterminate. Is it possible that such apparently fundamental differences between past and future are just accidents of the configuration of things?

This is astounding. But *this* is something that, I believe, we can untangle.

4

Think of two tanks full of water, with a short interconnecting channel that can be opened or closed.

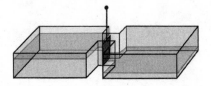

If the bulkhead is open, the water in the two tanks settles at the same level. This is an equilibrium state. Nothing happens if it is left alone. Everything is static, nothing distinguishes past from future. A film of the water, run backward, is indistinguishable from one played as it was filmed.

Close the bulkhead and add water to one of the tanks. The level of the water is now higher in that tank. Each tank is in equilibrium in itself, but the two tanks together are not in equilibrium with each other. There is a disequilibrium, maintained by the closed bulkhead, that prevents diffusion. In this case, too, everything is static, and nothing distinguishes past from future. A film of the water projected backward is still indistinguishable from one projected the right way.

Now think about what happens if the bulkhead is briefly opened. Some water enters the connecting section, flows toward the shallower pool and generates a wave that spreads in the second tank.

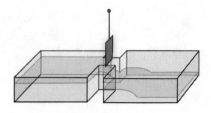

The wave rebounds from the sides of the walls, disperses, and after a while the water is still again. The height of the water in the two containers is a little more balanced.

All of this forms part of our everyday experience. The energy of the wave liberated by opening the bulkhead is called "free energy." Free energy gets consumed: when the wave subsides, that free energy no longer exists; it has "dissipated."

It has been dispersed among the molecules of water. It has spread out in the disordered motion of the water molecules. The free energy has dissipated into heat.

The interesting phase in this process is the intermediate one—after the opening of the bulkhead but before the reestablishment of calm. What happens in *this* (and only in this) time interval is oriented in time: if I film it and run the film backward, I see something absurd—water that becomes agitated by itself, for no reason, forms into a wave, rushes into the connecting space and gathers at the bulkhead an instant before it closes—something that could not happen in reality.

The movement of water to the less full tank is an *irreversible* phenomenon. Like the breaking of an egg that cannot put itself together again. Prior to the opening of the bulkhead, everything is reversible. In the intermediate phase there is irreversibility.

This irreversibility is generated by three ingredients: the initial disequilibrium (water at different levels in the two tanks; something that had maintained this disequilibrium for a long time); the bulkhead; and the fact that it takes time to reach a new equilibrium. These three conditions—initial disequilibrium, isolated systems that occasionally interact, long periods of equilibrium—are present everywhere in the universe that we find ourselves living in:

1. In the past the universe was very compressed, and this was a situation of disequilibrium.

2. The universe is full of disequilibria maintained by "bulkheads." Hydrogen and helium, for example, are in disequilibrium, like the tanks of water. The "bulkhead" that prevents their equilibrium is the fact that the transformation does not happen in the cold. Occasionally a large cloud of hydrogen is compressed by gravity and warms up as it is squashed; the temperature rises, and this opens up the possibility of its transformation into helium—the "bulkhead" that separates hydrogen from helium is opened. A star is born. Stars are the channel where a passage has been opened, like the connecting section that allows water to flow between the two tanks: hydrogen transforms into helium in the star and moves toward equilibrium. The process is irreversible, like the wave of water that throws itself into the less full tank.

3. The water in the tanks reaches equilibrium after a few minutes, but a star such as the Sun takes billions of years to burn. The wave of irreversibility it produces, like the wave that comes from the fuller tank, washes daily over the Earth, causing the innumerable irreversible processes that build the biosphere. Living beings such as ourselves are like the whirlpools generated by the waves of

water released by the opening of the bulkhead. We are the irreversible foam of free energy that was trapped in the disequilibrium between hydrogen and helium, freed by the Sun.

AND NOW WE CAN GET TO THE KEY POINT. LOOK AT the last figure, at the water pouring into the interconnection. Without any other information, you can tell that the bulkhead was recently opened. The wave *testifies* to the fact that something happened *before*—the opening of the bulkhead. Something in the present tells us of an event *in the past*.

Traces, memories, observations are all irreversible phenomena like this. For them to occur, it is enough that the three conditions I have listed are present: (1) there are systems in disequilibrium that (2) occasionally interact, and (3) the system that preserves the trace, the memory, the record must in turn remain away from equilibrium for a while.

The initial disequilibrium of the past is the reason the present has traces *of the past*. The formation of every trace is nothing other than an intermediate step toward equilibrium.[31] If the present has traces of the past, it is due solely to the disequilibrium of that past.

It is for this reason that we remember the past and not the future—because of the disequilibrium in the past. We know the past because there are traces of it in the present—in our memories, for example. To say that the past is determined is to say no more than that we have many traces of it. It is not a direction intrinsic to time that makes the past knowable, determined: what we call the past is how things were arranged at one point in time. It is the disequilibrium of the past—only that—that gives rise to traces.[32]

A meteorite that falls on the Moon carries *free energy* with it. A crater is the trace that it leaves until the incessant unraveling of things erases it. In this intermediate phase the crater is a *trace* of the impact, a *memory* of it. Traces exist in this intermediate period. A crater is like the wave in the tank, only over a much longer period of time. The same goes for a photograph, for the memory in our brain. It exists thanks to the fact that free energy has arrived in a system (the camera film, our brain) from another system that was not in equilibrium with it, and the fact that it takes time for equilibrium to be reestablished.

The reason we remember the past and not the future is entirely due to the fact that the universe was further from equilibrium at one point in the past than it is now.

If a system then reaches complete equilibrium, there are no more traces, no more memory, nothing to distinguish the past from the future. Sooner or later, every memory vanishes, canceled through the wear and tear of time. Sooner or later, of our proud civilizations, of everything that we have understood, of the words in books such as this one, of our controversies and of our desperate passions and loves . . . no trace will remain.

5

The second phenomenon that seems to contradict the fundamental similarity of past and future, a phenomenon that concerns us even more than memory, is the fact that we can *choose* the future, but we cannot choose the past. We decide. When we make a decision, we consider the pros and cons, assess the available information, consult our memory, assess our objectives, take our values into account, weigh our motivations, our desires, our deepest ethical convictions. And so on. And at the end of this process, we decide: "Yes, all things considered, I'm going to get the bar of chocolate from the cupboard."

Making a decision can be a complex process. A computer that plays chess and "thinks long about it" before moving does the same thing that we do, albeit in a less complex way. "Making a decision" is the name that we give to this complicated process that takes place between our neurons before an action. There is nothing strange about this: the world is full of complicated processes. But there is a radical aspect in decisions: we can decide "freely." Whether at the end of a troubled process of evaluation, or intuitively and without thinking, it is *we* who decide spontaneously, in a way that cannot be anticipated. The world can evolve, following this free decision of ours, toward different futures. After all, we might not have eaten the bar of chocolate (or so we say—after having eaten it). We can "freely" decide—but only the future, not the past.

Where does *this* asymmetry of time come from?

The answer is the same, once again: it is the result of the disequilibrium of the world we live in. A decision, as well, is an irreversible step toward equilibrium.[33] The freedom of choice is real, but it regards the *macroscopic* description of what happens, not the microscopic one. It is the *macro*-story that branches. This is possible because different *macro*-futures are compatible with the same *macro*-past. And this, in turn, is possible precisely because many different *micro*-pasts correspond to the *macro*-past.

Freedom, *that we search, and which is so dear to us,** is real. But as Spinoza clarified in the seventeenth century, freedom is our way of designating the fact that we are not able to fully

---

* *Purgatory*, I. With similar words Virgil describes Dante's quest to the guardian of Purgatory.

reconstruct what happens in the decision making, to predict what we will decide. Spinoza writes: "Men feel free, because they are aware of their choices and their wishes, but they ignore the causes that lead them to will and to choose, and do not give the slightest attention to these causes."[34] And again: "Men think that of their own free will they can do something or refrain from doing it, opinions that consist only of this, that . . . they ignore the causes that make them act as they do."[35]

THERE ARE THOSE, STRANGELY, WHO ARE DISTURBED by this understanding of what freedom means. I think they are making a mistake. The old fisherman's mistake.[36]

once upon a time there was an old fisherman who loved the sunset very much. the horizon explodes with fiery colors, the sun descends majestically and slowly plunges into the ocean, the sky is colored with *the sweet color of oriental sapphire*,* and one by one the stars appear. then one day a visitor from the city says to the old man: "you know, the sun does not sink into the sea. the sun is still out there, always shining. the show you see is just a trick of perspective, due to the rotation of the planet that we're standing on." the old fisherman is astonished. he trusts the man from the city. he begins

---

* *Purgatory*, I. The color of the sky, seen by Dante after emerging from Hell.

to be troubled. the sunset is an illusion—the man had said—therefore it is not real. for years the fisherman had been absorbed and fascinated by an unreal event. he had been tricked his entire life. if the sunset is an illusion, he now thinks, it cannot be relied on. he must learn to do without sunsets. he attempts to do so; it is a disaster. he does not know when to go to sleep; in the evening he does not expect the night, and when the sunset comes, he repeats, "it's an illusion, it's not true, there's no such thing as a sunset, the sun does not sink into the sea. the sun always shines. i have to take reality seriously. i do not have to go to sleep." he never sleeps again; he ends up insane . . .

The old man was obviously committing an error, but a subtle one. The question that disturbed him was whether the sunset was real or illusory. The sun does not sink into the sea. But denying the reality of the sunset leads to dramatic and senseless deductions. Where does the mistake lie?

It lies in confusion about the meaning of "sunset." The old man grew up with an idea of what a sunset is. It is the sun sinking into the waters of the ocean. When he learns that the sun does *not* sink into the sea, he concludes that there is no sunset.

But we, who know our Copernicus, still speak calmly about sunsets, even though we know that the sun does not move. We enjoy sunsets, we rely on them, and it does not cross our minds to say that there is no such thing.

We have readjusted our notion of "sunset." For us a sunset is real; it is the one that we have always seen. But it is no longer the sun sinking into the sea. It is, if we really want to think about it, what happens when the rotation of the Earth takes us away from the illuminated part. It is still the same sunset.

So why should we be disturbed to discover that past and future are only perspective phenomena? That our freedom is a macroscopic phenomenon that is not borne out at the microscopic level? This is the same as discovering that the sunset is not the sun sinking into the sea. It changes nothing in our lives. But it redefines what we mean by the flow of time: it is nothing other than a name given to a particular perspective on a peculiar arrangement of things . . .

THERE IS MORE, I THINK. TO DISCOVER THAT THE SUBtle logic that orients black holes is the same that orients our memory and our choices is to understand that we are part of the same global flux, the same eternal current.

All of the information in the macroscopic world arises from the dissipation of a disequilibrium in the past.[37] The information stored in every memory comes from the information implicit in past disequilibrium. The information created in every free choice is paid for by a decrease in disequilibrium, again from the disequilibrium of the past.

This brings us to a conclusion that seems to me extraordinary. Our neurons, our books, our computers, the DNA in our cells, the historical memory of an institution, the entire contents of the data on the internet, *my sweet guide, whose holy eyes were glowing as she smiled,*[*] the ultimate source of all the information of which life, culture, civilization are made, is none other than the disequilibrium of the universe in the past.[38]

The whole biosphere and the entirety of human culture are like the vortices of the wave created between the two tanks of water, the irreversible fall from a state of disequilibrium, delayed over billions of years, by the slowness of the phenomena of equilibration.

Furthermore, this is the only reason effects come *after* and not *before* causes. A cause is an intervention that leaves a trace, a memory—its effect. The relation between cause and effect is a step toward the equilibration of the world. The physics of causes and of effects is the same as the physics of traces and of memory. It is all about equilibration.[39] The direction of time is this equilibration of things. It is this journey toward equilibrium. It is an accidental phenomenon, due to a particular unbalanced state of affairs in the time we call past.[40]

It is a perspective phenomenon. Because it regards the

---

[*] *Paradise*, III. Beatrice, obviously.

macroscopic description of the world, it depends on the macroscopic variables used to describe it. Yet perspective phenomena can be magnificent. The daily rotation of the sun, the moon and the other stars around us is a perspective phenomenon—the stars and the sun do not move, but the rotation of the heavens is no less magnificent for that.

The same applies for the magnificence of the cosmic flow of time.

IN A UNIVERSE IN EQUILIBRIUM, AS IN THE TANK AFTER the wave has subsided, there is no phenomenon allowing us to distinguish past from future. We could not say in which direction time flows.

But there would be a far more radical consequence for us in such a universe: our very thoughts could not exist. We would not be able to make observations, or to reason, because to think is to dissipate energy. We would not have senses, because our senses record things—in other words, they produce memories. They would not work in a situation of equilibrium. We would not be able to hear music, since music exists in our heads only because we can *remember* preceding notes. In a universe in equilibrium, we would simply not exist as thinking and feeling beings.

It is because disequilibrium is so necessary to thinking that it is so natural to us to think of time as directional, and so difficult to accept the idea that the orientation of time is not fundamental. The time of our thinking is directional because our thinking is itself an irreversible phenomenon. Because we ourselves are irreversible phenomena. We are children of time.

In a naturalized version of Kant, we can say that an arrow of time—that is, the three conditions outlined above: disequilibrium, systems' separation and long relaxation times—is a necessary a priori condition for consciousness, because knowledge is a natural phenomenon in natural beings such as ourselves, whose sensibility and thoughts are a macroscopic phenomenon that depends precisely on the macroscopic directionality of time.

This, in the end, is the answer to the question of why it is so difficult for us to accept the idea of a nature that is not oriented in time—because our very thinking is itself a child of the orientation of time. It is itself one of the products of the initial disequilibrium.

We always make the mistake of thinking of ourselves as different from the world around us, of thinking that we are looking at it from the outside. We forget that we are like other things—that we, too, are like the things we look at.

For this reason, every investigation of things ends up closely involving ourselves.

When we seek to understand white holes, we do so not as pure reason, not as part of a world different from the objects that we are trying to understand. We are processes guided by the same stars.

PERHAPS THIS IS THE REASON WE ARE INTERESTED IN what happens at the end of a fall into a black hole . . . come to think of it, it is also perhaps the real reason i write. or better: the reason i write and rewrite these pages, composed in layers and continually shuffled and reshuffled . . . the order of the words has little to do with the jumbled order in which they were born (what i am writing now is the fifth revision). the order of time has always something of a reconstruction about it. the flow of reality is always more fluid than any of our frantic attempts to capture it might lead us to believe. time is not the map of reality: it is a kind of memory storage device . . .

to study something is to enter into a relation with that thing, to form correlations that allow us to represent, simplify and predict how that thing, that process, will unfold.

to understand is to identify with the thing understood, to construct a parallel between something in the structure of our synapses and the structure of the object in which we are interested. knowledge is a correlation between two parts of nature. understanding is a more abstract but also a more intimate commonality between our minds and phenomena.

this interweaving of correlations—between the endless

richness of our individual and collective memory, and the fabulous richness of the structure of reality—is itself an indirect product of the equilibration of things in time.

we, creatures of thought and of emotion, are this interweaving that is formed at the macroscopic level between ourselves and the world. we are not just social beings who live on relationships with other human beings, and biochemical organisms that burn free energy from the sun, in common with the rest of the biosphere. we are also animals endowed with neurons that are interwoven, thanks to these correlations, with other parts of reality.

we are as curious as cats about everything. even about white holes. it is in our nature to "go and see." to call this "curiosity," therefore, is perhaps reductive. it is our natural way of going toward things, because things are what we are. they are our sisters.

the emotion of discovery, the hours spent discussing and thinking, the joyful lightness of that day with hal . . . all this is not just curiosity. it is a strange, uncertain desire to get closer to things. *along the solitary plain we go . . .*

in the end, it seems to me, the real purpose of language is not to communicate. it is to get close to things, to be in relation with them.

when we speak with friends, with the people we love, do we really speak in order to tell them something? don't we really use the excuse of wanting to tell them something, in order to be able to speak with them?

when dante in *paradise* interrogates beatrice on questions of doctrine, is he really motivated by doctrinal questions? rather, isn't the objective to get to the point where *Beatrice gazes upon me with her eyes Full of the sparks of love, and so divine, That, overcome my power, I turned my back And almost lost myself with eyes downcast*?*

the same, i think, goes for the world. to study space, time, black holes and white holes is a way of us being in relation to reality. a reality that is not "it," but "you"—as lyric poets have it when addressing the moon. in *the jungle book*, all the animals give each other the cry of mutual recognition: "we be of one blood, ye and i."

i happen to think that we should always address the universe as "you," to understand it and to understand ourselves. a "you" that recognizes our shared identity with things, that says: we be of one blood, ye and i. whenever it is a damp, drizzly november in our soul, then, it is high time to get to sea and quietly board a ship that takes us to the world.

very many years ago, traveling alone in india, i found myself crammed and buffeted for hours on end in a ramshackle bus crowded beyond capacity with human and animal fellow

---

* *Paradise*, IV.

passengers, crawling in the torrid heat through a countryside that seemed endless. pressed against me and equally tossed about, there was a small, shy-looking boy dressed in a white tunic. after a long time, he cautiously ventured to ask me a question. the question, without preamble, was what was my path toward god. naturally i did not have an answer. perhaps today, though, decades later, i would have something with which to reply.

according to a sioux elder, it is said, the meaning of life is to address everything that we encounter with a song.

this is my song to white holes.

6

We have the complete picture. A large cloud of hydrogen sailing through cosmic spaces begins to get denser under the weight of its own gravity. As it contracts, it heats up until it ignites and becomes a star. The hydrogen burns for billions of years until it is wholly transformed into helium and ashes. Gravity becomes irresistible and the star collapses into a black hole. Other black holes can perhaps form in the inferno of the primordial universe, when the fluctuations and temperature of everything are violently extreme.

However formed, the matter sinks and rapidly reaches the center. Here the quantum structure of space and of time prevents it from being crushed any further. It has become a Planck star. It then "bounces" and begins to explode.

Around it, inside the hole, space itself completes a quantum leap and its geometry rearranges itself, transforming like Gandalf from gray to white.

This process of transition is of the same kind that produced the Big Bang, possibly from the collapse of a previous universe: space and time dissolve and reform. It's a process that is outside of time and outside of space, and yet described by the equations of quantum gravity.

In a white hole, everything that falls then flies upward. In the end, everything that has entered comes out entirely from the white horizon and again returns to seeing *the sun and the other stars*.*

Seen from the outside, the process in its entirety lasts for an extremely long time. Even for billions of years or more. A black hole takes eons to evaporate,[41] a white hole an even longer time to dissipate,[42] to let out all of the information and what little residual energy remains, until the long, happy life of this extraordinary process is complete.

Long, but nevertheless finite, as finite are the lives of us all, of every living organism, every star, every galaxy, of all stories, in this universe of joy and pain. Not even white holes last forever.

But "very long" is the time of the process as witnessed by

---

* *Paradise*, XXXIII. These are the last words of *The Divine Comedy*. Each of the three *cantiche* of the *Comedy—Inferno, Purgatory* and *Paradise*—ends with the word "stars."

those who are *outside the hole*, those who see the collapse of a star and who wait for the black hole to evaporate, to transform into a white hole, and for however much is inside the white hole to slowly escape until the dissipation of the horizon.

Whoever enters the horizon instead (as we did, together with the matter that forms the hole by collapsing), or does so at any subsequent moment, would arrive at the quantum zone in a fraction of a second—or at the most in a few hours, if the star was truly gigantic—crossing it in a heartbeat, and in an equally short space of time coming out of the white hole's horizon, finding herself transported into a very distant future, in just a brief moment of *her* time.

A few instants inside are billions of years outside. Such starkly different time perspectives coexist in our universe. Our usual intuition of the long communal life of the universe is distorted. Gravity warps time more than we can imagine. The entire process of the life of a black hole and a white hole is like a shortcut—lasting an instant—to an immensely distant future.

This, in the end, is a Planck star: a shortcut to the future. A way to hide safely for a moment, while outside eons of time flow slowly by.

And yet, even this is only the dissipation of concentrated free energy—a small chapter in a global growth of entropy. On the one hand, white holes disrupt our sense of time;

on the other, they show us once again the vastness of the great river that is the dissipation toward equilibrium: Rilke's *eternal current, that whirls all the ages through both realms, overwhelming them in both.*

For anyone outside, the white hole that remains over a long time is a small, very stable object that weakly irradiates its minute residue of energy. Inside there is still a vast world; from outside it behaves like a simple, extremely small mass, with completely normal gravity.

A mass of what size exactly? No smaller than the Planck mass, because a horizon with Planck mass has the dimensions of a Planck area, and the granularity of space prevents the existence of anything smaller. But not much bigger either, because a large white hole is unstable and would turn black.[43] A Planck mass is the mass of a small hair. A speck of dust.

A white hole in the sky is like a floating speck of dust.

Unlike a speck of dust, it does not have electrical features, and therefore does not interact with light. It cannot be seen. It only has its extremely weak gravitational force.

Whether in the primordial universe or in a phase preceding the Big Bang, many black holes could have formed that are now already evaporated. It is possible that right now they float in the sky in their millions—invisible grains, of a few fractions of a gram.

❦

ARE THEY REALLY OUT THERE?

who knows. hal and i would love it if it were the case. as for that rapid first look: true love stories only open, never close. the story that i have told and retold in the writing and rewriting of these lines is hardly concluded. it is unfolding. as ever, we look toward the mystery, trying to read the signs peering through the dark.

perhaps—as was the case for decades with the black hole at the center of the Milky Way, the hiss of which was heard by millions of americans on the evening of may 15, 1933, without anyone understanding what it was—these minuscule white holes in the heavens might already have been revealed long ago without anyone recognizing them yet. after all, astronomers have long since observed that the universe is teeming with a mysterious invisible dust that reveals itself only through its gravity and that they have called "dark matter."

this dark matter might be made up, perhaps in part, of billions upon billions of these tiny, delicate white holes that reverse the time of black holes, and float lightly throughout the universe, like dragonflies . . .

London, Ontario; Marseille; Verona

2020–2022

NOTES

1. Vladimir Mayakovsky, "At the Top of My Voice," *The Bedbug and Selected Poetry*, trans. Max Hayward and George Reavey (Cleveland: Meridian Books, 1960).

2. David Finkelstein, "Past-Future Asymmetry of the Gravitational Field of a Point Particle," *Physical Review* 110 (1958), 965–67.

3. David Finkelstein, "MELENCOLIA I: The Physics of Albrecht Duerer," https://doi.org/10.48550/arXiv.physics/0602185. Published also as *The Melencolia Manifesto* (San Rafael, CA: Morgan & Claypool, 2016).

4. I describe the interior of Schwarzschild geometry using the foliation that maximizes the volume of the equal-time surfaces. For technical details, see Marios Christodoulou and Carlo Rovelli, "How Big Is a Black Hole?," *Physical Review D* 91 (2015), 640–46.

5. The drawing is lacking a dimension: the circles represent spheres.

6. Linji Yixuan, *The Record of Linji*, trans. Ruth Fuller Sasaki (Honolulu: University of Hawai'i Press, 2009).

7. We are using the definition in note 4 above.

8. The Planck length is extremely small, 10^{-33} centimeters, but the radius of the cylinder does not need to be this small to be the quantum zone. The curvature of the black hole is in the order of its mass

divided by the cube of the radius ($R \sim M/r^3$), so if the mass is large enough the radius can be large.

9. See, for instance, Fenrong Liu, "New Perspectives on Mohist Logic," *Journal of Chinese Philosophy* 37, no. 4 (2010), 605–21.

10. Carlo Rovelli and Francesca Vidotto, "Planck Stars," *International Journal of Modern Physics D* 23 (2014), 1420–26.

11. To be precise, it is not its volume but rather its density that has reached the Planck scale.

12. Carlo Rovelli and Lee Smolin, "Spin Networks and Quantum Gravity," *Physical Review D* 53 (1995), 5743–59, and "Discreteness of Area and Volume in Quantum Gravity," *Nuclear Physics B* 442 (1995), 593–619.

13. Spin labels the irreducible representations of *SU(2)*, the (covering group of the) group of the rotations.

14. As related to me by Roger Penrose.

15. See note 10.

16. Changing the time sign changes the velocity sign (first derivative), not the acceleration sign (second derivative), which remains attractive.

17. A highly astute reader may be doubtful, thinking, "But it's *improbable* . . ." I have dedicated several pages to this point. For the moment I am only talking about possibility, not probability.

18. This idea was already suggested in Rovelli and Vidotto, "Planck Stars," cited in note 10 above.

19. Essentially: changing coordinates.

20. Hal Haggard and Carlo Rovelli, "Black Hole Fireworks: Quantum-Gravity Effects outside the Horizon Spark Black to White Hole Tunneling," *Physical Review D* 92 (2015), 104–20, https://arxiv.org/abs/1407.0989.

21. Stephen Hawking, "Black Hole Explosions," *Nature* 248 (1974), 30–31.

22. Dissipation and entropy increase.

23. There may be others: Alejandro Perez, in Marseille, for instance, is

studying the possibility that there is dissipation toward Planck scale geometric degrees of freedom.

24. Entropy can be computed from the relation between energy and temperature.

25. Andrew Strominger and Cumrun Vafa, "Microscopic Origin of the Bekenstein-Hawking Entropy," *Physics Letters B* 379 (1996), 99–104; Carlo Rovelli, "Black Hole Entropy from Loop Quantum Gravity," *Physical Review Letters* 77 (1996), 3288–91.

26. Entropy is proportional to the area of the horizon, and the number of possible states is determined by the entropy.

27. Juan Maldacena, "Recent Progress on the Black Hole Information Paradox," paper presented at Strings 2020, Cape Town, South Africa, https://indico.cern.ch/event/929434/contributions/3913390 /attachments/2069777/3474397/Maldacena.pdf.

28. The information paradox is born out of the idea that the total number of states of a black hole is measured by the Bekenstein-Hawking entropy, hence by the area of the horizon. This is the dogma, which is an extreme version of holography (weaker versions of holography exist). It follows from the dogma that evaporation reduces the number of states: at Page time, there are no longer enough states to purify the Hawking radiation. Von Neumann entropy begins to decrease, giving rise to Page's curve. If so, there must be a mechanism that allows information to exit. Now, this argument is based upon two erroneous assumptions. The first is that the von Neumann entropy is always less than the thermodynamic entropy. This is true only for ergodic systems, and a black hole certainly isn't one, due to its causal structure, which does not allow the equipartition of energy between the interior and the horizon. The causally disconnected part of the system continues to contribute to von Neumann entropy—from entanglement formed in the past— but not to thermodynamic entropy. When the horizon evaporates, its *thermodynamic* entropy lowers, but its *von Neumann* entropy does not, allowing information to remain inside, without contributing

to the thermodynamics of the hole as seen from the exterior. The second erroneous assumption is that the horizon is an *event* horizon. It is not, precisely because of the black-to-white transition. The horizon is an apparent horizon: whether it is an event horizon or not depends upon quantum gravity, because the external curvature becomes Planckian *before* evaporation is ended. The argument that there is a problem with unitarity Page time depends on the (wrong) assumption that the horizon is an event horizon, hence on (mistaken) hypotheses about quantum gravity. The calculation of the number of states of strings regards a stationary black hole exterior, hence only event horizons. It regards the number of states *distinguishable from the outside*—where the observables are located, in this formulation. These are states of the horizon, not of what is inside. The information stays in the black hole. It exits when the hole has turned into a white hole capable of existing for a long time.

29. Perhaps (it is not clear to me) it can even happen before this, when the horizon is still quite large.

30. A very small (Planck mass) white hole is stable, thanks to quantum gravity.

31. Producing entropy.

32. For a detailed discussion, see Carlo Rovelli, "Memory and Entropy," *Entropy* 24, no. 8 (2022), 1022, https://arxiv.org/abs/2003.06687.

33. It necessarily produces entropy.

34. Spinoza, *Ethics*, appendix to Part I.

35. Spinoza, *Ethics*, Part II, Proposition 35.

36. Carlo Rovelli, "The Old Fisherman's Mistake," *Metaphilosophy* 53, no. 5 (2022), https://onlinelibrary.wiley.com/doi/10.1111/meta.12589.

37. Disequilibrium is information, because the greater the equilibrium, the greater the number of *micro*-states, and the less the information contained in the macroscopic state.

38. The low entropy of the past is the ultimate source of all the information contained in every trace or memory.

39. The distinction between causes and effects has no meaning in the microscopic description of phenomena. At the microscopic level of things there are regularities, physical laws and probability—and these notions do not distinguish between past and future. The distinction between past and future is a property of the history of the universe from the variables that we call macroscopic. It is only for this reason that we can speak about causes.

40. See Carlo Rovelli, "Back to Reichenbach," 2022, http://philsci-archive.pitt.edu/20148/.

41. A time in the order of m^3 in Planck units, where m is the *initial* mass of the black hole.

42. A time in the order of m^4 in Planck units.

43. Macroscopic white holes are unstable. White holes of Planck mass are stabilized by quantum gravity. Carlo Rovelli and Francesca Vidotto, "Small Black/White Hole Stability and Dark Matter," *Universe* 4 (2018), 127.

ILLUSTRATION CREDITS

INDEX

Italicized page numbers indicate material in photographs or illustrations.